Sabrina Hupp

Modulation of actin dynamics by the S.pneumoniae toxin pneumolysin

Sabrina Hupp

Modulation of actin dynamics by the S.pneumoniae toxin pneumolysin

A novel mechanism beyond pore formation

Südwestdeutscher Verlag für Hochschulschriften

Impressum/Imprint (nur für Deutschland/only for Germany)
Bibliografische Information der Deutschen Nationalbibliothek: Die Deutsche Nationalbibliothek verzeichnet diese Publikation in der Deutschen Nationalbibliografie; detaillierte bibliografische Daten sind im Internet über http://dnb.d-nb.de abrufbar.
Alle in diesem Buch genannten Marken und Produktnamen unterliegen warenzeichen-, marken- oder patentrechtlichem Schutz bzw. sind Warenzeichen oder eingetragene Warenzeichen der jeweiligen Inhaber. Die Wiedergabe von Marken, Produktnamen, Gebrauchsnamen, Handelsnamen, Warenbezeichnungen u.s.w. in diesem Werk berechtigt auch ohne besondere Kennzeichnung nicht zu der Annahme, dass solche Namen im Sinne der Warenzeichen- und Markenschutzgesetzgebung als frei zu betrachten wären und daher von jedermann benutzt werden dürften.

Coverbild: www.ingimage.com

Verlag: Südwestdeutscher Verlag für Hochschulschriften GmbH & Co. KG
Heinrich-Böcking-Str. 6-8, 66121 Saarbrücken, Deutschland
Telefon +49 681 37 20 271-1, Telefax +49 681 37 20 271-0
Email: info@svh-verlag.de

Approved by: Würzburg, Universität, Dissertation, 2012

Herstellung in Deutschland (siehe letzte Seite)
ISBN: 978-3-8381-0991-6

Imprint (only for USA, GB)
Bibliographic information published by the Deutsche Nationalbibliothek: The Deutsche Nationalbibliothek lists this publication in the Deutsche Nationalbibliografie; detailed bibliographic data are available in the Internet at http://dnb.d-nb.de.
Any brand names and product names mentioned in this book are subject to trademark, brand or patent protection and are trademarks or registered trademarks of their respective holders. The use of brand names, product names, common names, trade names, product descriptions etc. even without a particular marking in this works is in no way to be construed to mean that such names may be regarded as unrestricted in respect of trademark and brand protection legislation and could thus be used by anyone.

Cover image: www.ingimage.com

Publisher: Südwestdeutscher Verlag für Hochschulschriften GmbH & Co. KG
Heinrich-Böcking-Str. 6-8, 66121 Saarbrücken, Germany
Phone +49 681 37 20 271-1, Fax +49 681 37 20 271-0
Email: info@svh-verlag.de

Printed in the U.S.A.
Printed in the U.K. by (see last page)
ISBN: 978-3-8381-0991-6

Copyright © 2012 by the author and Südwestdeutscher Verlag für Hochschulschriften GmbH & Co. KG and licensors
All rights reserved. Saarbrücken 2012

To

my daughter Liliane,

Alex Bohl, the best technician in the world,

and

my bredrins and sistren in the motherland

"I nah have education. I have inspiration.
If I was educated, I'd be a damn fool."
(Bob Marley)

Index

Figures V
Tables VI

PUBLICATIONS.. 1

ABSTRACT... 3
ENGLISH... 3
DEUTSCH 6

1 INTRODUCTION 9
1.1 *STREPTOCOCCUS PNEUMONIAE*.. 9
1.2 BACTERIAL MENINGITIS.. 10
1.2.1 Epidemiology.. 10
1.2.2 Prevention and treatment 12
1.3 PNEUMOLYSIN - A BACTERIAL PROTEIN TOXIN 12
1.3.1 Pore-forming toxins 12
1.3.2 Pneumolysin - a member of the CDC family 13
1.3.3 Role in pathogenicity.. 15
1.4 THE CYTOSKELETON 17
1.4.1 General properties... 17
1.4.2 Actin - treadmilling - nucleation... 20
1.4.3 The actin cortex and active cell movement 22
1.4.4 Actin binding proteins (ABPs) 23
1.4.4.1 The actin-related protein complex (Arp2/3) 24
1.4.4.2 Gelsolin... 25
1.5 HOW BACTERIAL TOXINS INTERVENE WITH CELLULAR FUNCTIONS..26
1.5.1 Toxin strategies... 26
1.5.2 Small GTPases.. 28
1.5.3 Toxin effects on small GTPases 30
1.6 THE BRAIN 32

1.6.1	Brain anatomy	32
1.6.2	Neuroglia	34
1.6.3	Astroglia	34
1.6.3.1	Astrocytic features	34
1.6.3.2	Astrocyte - neuron interaction	36
1.6.3.3	Astrogliosis	37
1.6.3.4	The blood-brain barrier	38
1.7	BASICS OF THIS STUDY	39
2	METHODS	41
1.8	PNEUMOLYSIN PREPARATION	41
1.9	PNEUMOLYSIN CONCENTRATION AND LYTIC CAPACITY DETERMINATION	41
1.10	LABELING OF PLY WITH ATTO488	42
1.11	PREPARATION OF CELLS	43
1.12	CELL TREATMENT	43
1.13	IMMUNOCYTOCHEMISTRY	44
1.14	TRANSFECTION OF CELLS	45
1.15	BACULOVIRUS TRANSDUCTION	46
1.16	MICROSCOPY	46
1.16.1	Bright field microscopy	46
1.16.2	Fluorescent microscopy	47
1.16.3	Live cell imaging / time lapse imaging	47
1.16.4	Confocal microscopy	48
1.17	PROTEIN BIOCHEMISTRY	48

1.17.1	Small GTPase activation pull down assays	48
1.17.2	Cytoskeleton Isolation Assays	49
1.17.3	Preparation of G-actin and F-actin	50
1.17.4	Actin binding assays	51
1.17.5	Enzyme-linked sorbent assays (ELSAs)	51
1.17.6	Actin-pyrene polymerization assays	52
1.17.7	Imaging of actin-rhodamine filaments	54
1.17.8	Far Western Blotting (Overlay Blot)	55
1.17.9	FRET experiments	56
1.18	GIANT UNILAMELLAR VESICLE (GUV) APPROACH	56
1.19	CELL BORDER TRACKING ASSAY	58
1.20	PREPARATION AND RESEALING OF ERYTHROCYTE GHOSTS	59
1.21	GRAPHICAL AND STATISTICAL ANALYSIS	60
1.22	EQUIPMENT AND MATERIALS	61
1.22.1	Pneumolysin and plasmids	61
1.22.2	Materials, animals, equipment and software	62
1.22.3	Buffers and solutions	67

2	RESULTS	71
2.1	DETERMINATION OF THE ASTROGLIAL NATURE OF CELLS	71
2.2	DEFINITION OF SUB-LYTIC CONCENTRATIONS OF PNEUMOLYSIN (PLY)	72
2.3	INVESTIGATION OF EFFECTS OF SUB-LYTIC PLY CONCENTRATIONS ON ASTROCYTES AND THE CYTOSKELETON	74
2.4	INVOLVEMENT OF THE SMALL GTPASES RAC1 AND RHOA IN THE EFFECTS INDUCED BY PNEUMOLYSIN	81
2.5	APPLICATION OF GIANT UNILAMELLAR VESICLES (GUVS) FOR INVESTIGATION OF TRANSMEMBRANOUS PLY EFFECTS	83
2.6	INVESTIGATION OF PLY-ACTIN INTERACTION	85

2.6.1	Spin-down assays	85
2.6.2	Actin-pyrene polymerization assays	87
2.6.3	Far Western Blotting	88
2.6.4	Enzyme-linked sorbent assays (ELSAs)	89
2.6.5	Immunocytochemistry	90
2.7	INVESTIGATION OF PLY-ARP INTERACTION	91
2.7.1	Enzyme-linked sorbent assays (ELSAs)	91
2.7.2	Immunocytochemistry	92
2.7.3	Actin-pyrene polymerization assays	92
2.7.4	Imaging of actin-rhodamine (TRITC-tagged)	94
2.8	INVESTIGATION OF STRUCTURAL CHARACTERISTICS OF PLY-ACTIN AND PLY-ARP2 INTERACTION	95
2.8.1	Enzyme-linked sorbent assays	95
2.8.2	Actin-pyrene polymerization assays	96
2.8.3	FRET (Fluorescence resonance energy transfer)	97
2.9	INFLUENCE OF THE ARP2/3 COMPLEX ON PORE FORMATION - APPLICATION OF ERYTHROCYTE GHOSTS	98
3	DISCUSSION	100
4	CONCLUDING REMARKS	116
5	PERSPECTIVES	120
Abbreviations		122
References		i

Figures

Fig. 1:	*S. pneumoniae* (electron microscopy).	9
Fig. 2:	The meninges.	10
Fig. 3:	Distribution of meningitis cases in 2002.	11
Fig. 4:	Protein configuration of pneumolysin monomer.	14
Fig. 5:	Pore-formation mechanism.	15
Fig. 6:	Schematic of the components of the cell cytoskeleton.	18
Fig. 7:	Actin treadmilling and nucleation.	19
Fig. 8:	Actin monomer and actin filament.	21
Fig. 9:	Active movement of animal cells mediated by the actin cortex.	23
Fig. 10:	The Arp2/3 complex.	24
Fig. 11:	Molecular mechanism of small GTPase activation and deactivation.	29
Fig. 12:	Mechanism of actin modification by Rac and Cdc42.	30
Fig. 13:	Anatomy of the brain.	33
Fig. 14:	Astroglia. Schematic of astrocytes within the neuronal network.	35
Fig. 15:	Co-localization of neurons and astrocytes.	37
Fig. 16:	Barrier sites in the CNS.	39
Fig. 17:	Magnetofection™ (schematic).	45
Fig. 18:	Baculovirus transduction with Organelle Lights™ (schematic).	46
Fig. 19:	Actin Spin Down Assay (schematic).	51
Fig. 20:	ELSA assay (schematic).	52
Fig. 21:	Time course of in vitro actin polymerization in the absence of effector proteins.	53
Fig. 22:	Nanion Vesicle Prep Pro®.	58
Fig. 23:	Displacement tracking of living cells.	59
Fig. 24:	Schematic of erythrocyte preparation.	60
Fig. 25:	Primary mouse astrocytes.	71
Fig. 26:	Lytic capacity of PLY.	72
Fig. 27:	Behavior of mock-treated astrocytes.	74
Fig. 28:	Effects of sub-lytic amounts of PLY (0.2 µg/ml) on primary astrocytes.	75
Fig. 29:	Phases of cell displacement.	76
Fig. 30:	Actin cortex collapse.	77
Fig. 31:	Development of focal adhesions.	78
Fig. 32:	Pre-treatment of astrocytes with cytochalasin D.	79
Fig. 33:	Actin changes induced by PLY challenge.	80
Fig. 34:	Displacement of astrocytes in calcium-free buffer.	81
Fig. 35:	Involvement of Rac and Rho in the cellular changes induced by PLY.	82
Fig. 36:	GUV approach.	84
Fig. 37:	Cytoskeleton Isolation Assay.	86
Fig. 38:	Actin binding and stabilization capacity of PLY.	86
Fig. 39:	Actin stabilization capacity of PLY.	88
Fig. 40:	Overlay blot.	89
Fig. 41:	Affinity of PLY to actin.	89

Fig. 42:	Actin bundle decoration by PLY.	90
Fig. 43:	Binding of PLY to the Arp complex and to Arp2.	91
Fig. 44:	Co-localization of PLY and Arp2.	92
Fig. 45:	Arp2/3-mediated actin polymerization.	93
Fig. 46:	Arp activation capacity of PLY.	94
Fig. 47:	Additive effects of PLY in the activation of Arp2/3 by VCA.	94
Fig. 48:	Fluorescent imaging of actin-rhodamine.	95
Fig. 49:	Structural properties of PLY interaction with G-actin.	95
Fig. 50:	Structural properties of PLY-Arp2 interaction.	96
Fig. 51:	Schematic of the unfolding domain of PLY.	97
Fig. 52:	Investigation of structural properties of PLY-Arp interaction.	97
Fig. 53:	FRET measurement between PLY and phalloidin.	98
Fig. 54:	Influence of Arp2/3 on the lytic capacity of PLY.	98
Fig. 55:	Schematic of actin nucleation activation.	118
Fig. 56:	Tissue penetration.	119

Tables

Table 1:	Pathogenicity of S. pneumoniae.	16
Table 2:	Bacterial toxins targeting Rho GTPases.	31
Table 3:	Bacterial toxins targeting actin.	32

PUBLICATIONS

Direct transmembrane interaction between actin and the pore-competent neurotoxin pneumolysin
Sabrina Hupp, Christina Förtsch, Carolin Wippel, Jiangtao Ma, Timothy J. Mitchell, and Asparouh I. Iliev
(submitted)

Astrocytic tissue remodeling by the meningitis neurotoxin pneumolysin facilitates pathogen tissue penetration and produces interstitial brain edema
Sabrina Hupp, Vera Heimeroth, Carolin Wippel, Christina Förtsch, Jiangtao Ma, Timothy J. Mitchell, Asparouh I. Iliev
Glia, 60, 137-146. 2012.

Changes in astrocyte shape induced by sublytic concentrations of the cholesterol-dependent cytolysin pneumolysin still require pore-forming capacity
Christina Förtsch*, **Sabrina Hupp***, Jiangtao Ma, Timothy J. Mitchell, Elke Maier, Roland Benz, and Asparouh I. Iliev
Toxins, 3, 43-62. 2011.

Extracellular calcium reduction strongly increases the lytic capacity of pneumolysin from *Streptococcus pneumoniae*
Carolin Wippel*, Christina Förtsch*, **Sabrina Hupp***, Elke Maier, Roland Benz, Jingtao Ma, Timothy J. Mitchell, and Asparouh I. Iliev
Journal of Infectious Diseases, 204, 930-936. 2011.

(*: these authors contributed equally)

ABSTRACT

English

Streptococcus pneumoniae is one of the major causes of bacterial meningitis, which mainly affects young infants in the developing countries of Africa, Asia (esp. India) and South America, and which has case fatality rates up to 50% in those regions. Bacterial meningitis comprises an infection of the meninges and the sub-meningeal cortex tissue of the brain, whereat the presence of pneumolysin (PLY), a major virulence factor of the pneumococcus, is prerequisite for the development of a severe outcome of the infection and associated tissue damage (e. g. apoptosis, brain edema, and ischemia). Pneumolysin belongs to the family of pore forming, cholesterol-dependent cytolysins (CDCs), bacterial protein toxins, which basically use membrane-cholesterol as receptor and oligomerize to big aggregates, which induce cell lysis and cell death by disturbance of membrane integrity.

Multiple recent studies, including this work, have revealed a new picture of pneumolysin, whose cell-related properties go far beyond membrane binding, pore formation and the induction of cell death and inflammatory responses. For a long time, it has been known that bacteria harm the tissues of their hosts in order to promote their own survival and proliferation. Many bacterial toxins aim to rather hijack cells than to kill them, by interacting with cellular components, such as the cytoskeleton or other endogenous proteins. This study was able to uncover a novel capacity of pneumolysin to interact with components of the actin machinery and to promote rapid, actin-dependent cell shape changes in primary astrocytes. The toxin was applied in disease-relevant concentrations, which were verified to be sub-lytic. These amounts of toxin induced a rapid actin cortex collapse in horizontal direction towards the cell core, whereat membrane integrity was preserved, indicating an actin severing function of pneumolysin, and being consistent with cell shrinkage, displacement, and blebbing observed in live cell imaging experiments.

ABSTRACT

In contrast to neuroblastoma cells, in which pneumolysin led to cytoskeleton remodeling and simultaneously to activation of Rac1 and RhoA, in primary astrocytes the cell shape changes were seen to be primarily independent of small GTPases. The level of activated Rac1 and RhoA did not increase at the early time points after toxin application, when the initial shape changes have been observed, but at later time points when the actin-dependent displacement of cells was slower and less severe, probably presenting the cell's attempt to re-establish proper cytoskeleton function.

A GUV (giant unilamellar vesicle) approach provided insight into the effects of pneumolysin in a biomimetic system, an environment, which is strictly biochemical, but still comprises cellular components, limited to the factors of interest (actin, Arp2/3, ATP, and Mg^{2+} on one side, and PLY on the other side). This approach was able to show that the wildtype-toxin, but not the Δ6 mutant (mutated in the unfolding domain, and thus non-porous), had the capacity to exhibit its functions through a membrane bilayer, meaning it was able to aggregate actin, which was located on the other side of the membrane, either via direct interaction with actin or in an Arp2/3 activating manner.

Taking a closer look at these two factors with the help of several different imaging and biochemical approaches, this work unveiled the capacity of pneumolysin to bind and interact both with actin and Arp2 of the Arp2/3 complex. Pneumolysin was capable to slightly stabilize actin in an actin-pyrene polymerization assay. The same experimental setup was applied to show that the toxin had the capacity to lead to actin polymerization through activation of the Arp2/3 complex. This effect was additionally confirmed with the help of fluorescent microscopy of rhodamine (TRITC)-tagged actin. Strongest Arp2/3 activation, and actin nucleation/polymerization is achieved by the VCA domain of the WASP family proteins. However, addition of PLY to the Arp2/3–VCA system led to an enhanced actin nucleation, suggesting a synergistic activation function of pneumolysin.

Hence, two different effects of pneumolysin on the actin cytoskeleton were observed. On the one hand an actin severing property, and on the other hand an actin stabilization property, both of which do not necessarily exclude each other.

Actin remodeling is a common feature of bacterial virulence strategies. This is the first time, however, that these properties were assigned to a toxin of the CDC family. Cytoskeletal dysfunction in astrocytes leads to dysfunction and unregulated movement of these cells, which, in context of bacterial meningitis, can favor bacterial penetration and spreading in the brain tissue, and thus comprises an additional role of pneumolysin as a virulence factor of *Streptococcus pneumonia* in the context of brain infection.

The aim of this study was to gain clarification of exact molecular mechanisms of cytoskeletal modifications. These findings in turn are prerequisite for the future development of efficient therapeutic approaches in order to better control the spreading and lethality of one major threat to a high number of people at risk of infection and little access to prevention.

Deutsch

Streptococcus pneumoniae gehört zur Gruppe der Pathogene, die bakterielle Meningitis verursachen, eine Infektion, die hauptsächlich bei Neugeborenen und Kleinkindern in den Entwicklungsländern von Afrika, Asien (Indien) und Südamerika auftritt, und in diesen Regionen Sterblichkeitsraten von bis zu 50% aufweist. Meningitis ist eine Infektion der Hirnhäute und dem sich direkt darunter befindlichen Cortex-Gewebe. Pneumolysin (PLY), ein Haupt-Pathogenitätsfaktor des sog. Pneumococcus, ist hauptsächlich verantwortlich für einen schweren Verlauf der Infektion und für Gewebeschädigungen, wie Apoptose, Hirnödemen und Ischämie.

Pneumolysin gehört zur Familie der Cholesterol-abhängigen Cytolysine (CDCs), bakteriellen Protein-Toxinen, die an Membran-Cholesterol binden, sich zu großen Aggregaten zusammenschließen und durch die Beeinträchtigung der Membranintegrität (Porenbildung) Zell-Lyse und Zelltod verursachen.

Zahlreiche neuere Studien, darunter auch diese Arbeit, haben ein neues Bild von Pneumolysin aufgezeigt, dessen Eigenschaften weit über die Membranbindung, die Poren-Bildung und die Induktion von Zelltod und inflammatorischen Prozessen hinausgehen. Es ist weithin bekannt, dass Bakterien das Gewebe ihres Wirts schädigen, um ihre eigene Vermehrung und ihre Ausbreitung zu begünstigen. In diesem Zusammenhang fungieren bakterielle Toxine als Pathogenitätsfaktoren, die mit zellulären Komponenten, wie dem Zytoskelett und anderen Zytosol-Proteinen interagieren, was allerdings bevorzugt zu Zellveränderungen, und seltener zum Zelltod führt.

Die vorliegende Arbeit konnte zeigen, dass Pneumolysin schnelle, und zum Teil gravierende, Aktin-abhängige Zellstruktur-Veränderungen in primären Astrozyten hervorruft. Hierbei wurde das Toxin in Konzentrationen appliziert, die im Liquor von Meningitis-Patienten detektiert werden können, und die zusätzlich als sub-lytisch für Astrozyten in Zellkultur verifiziert wurden. Diese Toxin-Mengen führten zu einem schnellen, horizontalen Aktinkortex-Kollaps, wobei die Membranintegrität erhalten blieb. Dies deutete auf eine „Severing"-Funktion (das Abtrennen oder Zerschneiden von Aktinfilamenten) von Pneumolysin hin, was mit den Beobachtungen übereinstimmt, die in Experimenten mit lebendigen Zellen gemacht wurden (Zellveränderungen, Zellbewegungen und „Blebbings").

Im Gegensatz zu Neuroblastoma Zellen, in denen Pneumolysin Zytoskelett-Veränderungen, und gleichzeitig die Aktivierung von Rac1 und RhoA verursachte, waren die Zell-Veränderungen bei Astrozyten primär unabhängig von der Aktivierung kleiner GTPasen. Obwohl gezeigt werden konnte, dass die Veränderungen vom Aktin-Zytosklett abhängig waren, war das Level an Rac1 und RhoA in den frühen Phasen nach der Toxin-Gabe nicht erhöht. Eine Aktivierung der GTPasen konnte dahingegen zu späteren Zeitpunkten detektiert werden, in denen die Zellbewegung abgeschwächt und verlangsamt war. Die späte Aktivierung kann als Reaktion der Zelle auf die vom Toxin ausgelösten Veränderungen gesehen werden, die zu einer Wiederherstellung der normalen Zytoskelett-Funktion führen soll.

GUV (giant unilamellar vesicle)-Experimente ermöglichen eine genauere Betrachtung der Pneumolysin-Effekte in einem biomimetischen, jedoch strikt biochemischen Ansatz, der alle zellulären Komponenten enthält, die untersucht werden sollen (Pneumolysin, Aktin, Arp2/3, ATP, und Mg^{2+}). Im GUV-System befand sich das Toxin im Inneren der Vesikel, und Aktin in der extra-vesikulären Suspension, einem Verhältnis genau umgekehrt zum zellullären System. Zusätzlich wurden Arp2/3 und ATP/Mg^{2+}, für die Aktin-Polymerisierung essentielle Faktoren, in der Aktin-Suspension zur Verfügung gestellt. Die GUV-Experimente konnten zeigen, dass Wildtyp-Pneumolysin, allerdings nicht seine Mutante Δ6-PLY (Mutation in der sog. unfolding domain, und deshalb nicht Poren-bildend), seine Effekte auf das Aktin-Zytoskelett durch die Membran-Barriere hindurch, in einer Membran-gebundenen Form ausüben kann. Aktin wurde an den Stellen höchster Toxinbindung aggregiert, was entweder über eine direkte Interaktion von PLY mit Aktin, oder über eine Aktivierung des Aktin-Effektors Arp2/3 durch Pneumolysin erklärt werden kann.

ABSTRACT

Weitere biochemische Ansätze (wie enzyme-linked sorbent assays, ELSAs) und Mikroskopie-Techniken (Immunocyto-Chemie) konnten beweisen, dass Pneumolysin sowohl mit Aktin, als auch mit Arp2 (einer Komponente des heptameren Arp2/3 Proteinkomplexes) direkt interagieren kann. Aktin-Pyren Experimente und Fluoreszenzmikroskopie (von TRITC-markiertem Aktin) wiesen auf eine Kapazität von Pneumolysin hin, Aktin direkt zu stabilisieren, und über die Aktivierung von Arp2/3 eine Aktin-Polymerisierung hervorrufen zu können. Der zelluläre Aktivator von Arp2/3 ist die VCA-Domäne von sog. WAS (Wiskott-Aldrich-syndrome)-Proteinen. VCA induziert eine Arp2/3-abhängige Aktin-Nukleierung, was zu einer schnellen Aktin-Polymerisierung führt. In diesem System erzeugte die Zugabe von Pneumolysin eine Verstärkung der Aktin-Nukleierung (synergistische Wirkung).

Demzufolge konnten in dieser Arbeit zwei abweichende Effekte von Pneumolysin auf das Aktin-Zytoskelett aufgezeigt werden. Einerseits zeigte das Toxin die Fähigkeit, den Aktinkortex von der Plasmamembran zu lösen („Severing"), und andererseits induzierte Pneumolysin eine Aktin-Stabilisierung, wobei sich beide Phänomene nicht grundsätzlich ausschließen.

Bakterielle Effektoren wirken häufig spezifisch auf das Aktin-Zytoskelett der Zellen im Gewebe des Wirts. In dieser Arbeit wurden diese Eigenschaften jedoch zum ersten Mal für ein Toxin der CDC-Familie vorgeschlagen. Eine Dysfunktion von Aktin in Astrozyten führt zu einer generellen Dysfunktion und zu einer unkoordinierten Bewegung der Zellen, was im Kontext von bakterieller Meningitis zu einer verstärkten bakteriellen Penetration und Ausbreitung im Hirngewebe des Wirtes führen kann, und somit zusätzlich die Rolle von Pneumolysin als Pathogenitätsfaktor von *Streptococcus pneumoniae* im Kontext von Infektionen des Gehirns bekräftigt.

Das Ziel dieser Arbeit, die molekularen Grundlagen der Pneumolysin-induzierten Zytoskelett-Veränderungen aufzuzeigen, ist Voraussetzung für die Entwicklung von effizienten therapeutischen Ansätzen, um die Ausbreitung und die Lethalität eines Pathogens zu kontrollieren, das im Besonderen eine große Anzahl von Menschen bedroht, die nur begrenzt Zugang zu Prävention und Behandlung haben.

1 INTRODUCTION

1.1 Streptococcus pneumoniae

Bacteria of the genus *Streptococcus* are Gram-positive, alpha-hemolytic, catalase-negative, anaerobic, but aero-tolerant. In Gram staining the organism appears as pairs of short chains of cocci (Fig. 1) [Kayser, 1998]. It was identified by Louis Pasteur in 1881 and given the name pneumococcus in the 1880s, as it was the most common cause of pneumonia [Marriott and Dockrell, 2006].

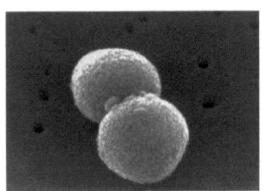

Fig. 1: S. pneumoniae (electron microscopy).
The Gram-positive pneumococcus appears as pairs of short chains of cocci in EM and Gram staining.
(Source: http://phil.cdc.gov, by R. Facklam, 2009; 2011)

S. pneumoniae is a commensal of the human upper airways, and becomes infectious only under immuno-suppressed conditions of the organism, and under conditions of gaining access to normally sterile spaces like the brain or bloodstream. However, colonization is the initial step in pathogenesis of all pneumococcal infections [Bogaert et al., 2004], whereat *S. pneumoniae* comprises a classical extracellular bacterium, as it is poorly opsonized and thus barely phagocytosed [Marriott and Dockrell, 2006].

Especially young children, elderly people, alcoholics, and patients with chronic disorders, such as HIV/AIDS, are susceptive to *S. pneumoniae* related conditions, such as pneumonia, bacteremia, sepsis, Otitis media and meningitis.

Pneumococcus expresses a variety of colonization and virulence factors (as shown in Table 1), including a polysaccharide capsule, surface proteins and enzymes, an autolysin, and the cytolysin pneumolysin (PLY) (reviewed in [Mitchell and Mitchell, 2010]). Up to date 90 different serotypes of *S. pneumoniae* are known, which are distinguished by capsular polysaccharide antigens [Henrichsen, 1995].

1.2 Bacterial meningitis

1.2.1 Epidemiology

Bacterial meningitis comprises an acute infection of the protective membranes covering the brain and spinal cord, the so-called meninges (Fig. 2) and the cortex tissue below the meninges. Brain tissue changes accompany the course of infections of the brain. They include brain swelling and perivascular and perimeningeal extracellular matrix digestion (most likely originating from the infiltrating leucocytes) [Kastenbauer and Pfister, 2003; Liu et al., 2008; Pfister et al., 1993].

Bacterial meningitis remains, despite of the availability of vaccination and antibiotic treatment, a threat to global health, causing death of estimated 170,000 infected worldwide per year (source: World Health Organization 2011).

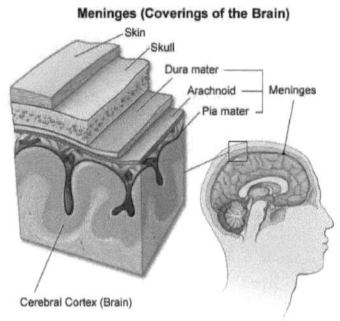

Fig. 2: The meninges.

The mammalian brain is covered by three protective layers. Meningitis is an acute infection of those layers and the closely associated cortex tissue directly below.

(Source: Yale School of Medicine; http://medicalcenter.osu.edu; 2011)

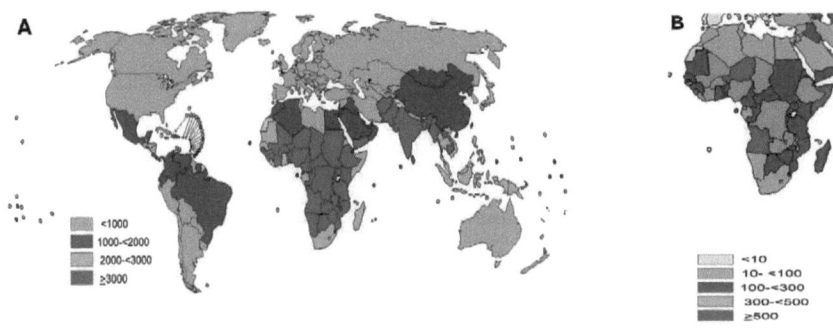

Fig. 3: Distribution of meningitis cases in 2002.
A. *Streptococcus pneumoniae incidence rate per 100,000 children under five years of age in 2000. In Africa most cases occur within the so called "meningitis belt", reaching from Senegal in the West to Ethiopia in the East.*
B. *Mortality rate per 100,000 children under five years of age due to Streptococcus pneumoniae, 2000 in Africa (modified). (Source: WHO, 2000; 2011)*

Large-scale epidemics occur especially in the African Sub-Saharan "meningitis belt" (Fig. 3) and widespread in India, but also in countries with high medical standards, pneumococcal infections come along with severe complications. Acute meningitis leads to death in about 33% of cases. More than 20% of surviving patients suffer from residual neurological symptoms (WHO 1997), such as mental retardation, learning disabilities and focal neurological deficits (e. g. hearing loss) (reviewed in [Bohr et al., 1984]). Under acute infectious conditions blood vessel inflammation, necrosis of the central nervous system (CNS), neuronal loss, general inflammation of brain tissue, and coma are common complications (further details in [Cairns and Russell, 1946]).

Most cases of meningitis are caused by the three bacterial species *Haemophilus influenzae* (Hib), *Streptococcus pneumoniae*, and *Neisseria meningitidis*. Since the introduction of Hib conjugate vaccines, *N. meningitidis* and *S. pneumoniae* have become the main causative agents of bacterial meningitis (WHO 2011). Latter is the primary factor for outbreak in children younger than one year of age (particularly during the neonatal period), and the second frequent cause in older children [Bingen et al., 2005]. Besides meningitis, *S. pneumoniae* causes pneumonia, Otitis media, and sepsis. All infections caused by pneumococcus comprise complications especially during early childhood. Recent estimates of child deaths caused by this pathogen range from 700,000 to 1 million every year worldwide [WHO, 1999, 2003, 2007].

1.2.2 Prevention and treatment

A 7-valent conjugate vaccine (PCV7), and a recently introduced 13-valent variant (PCV13) are available for prevention of invasive pneumococcal diseases (IPD), based on a reduction of nasopharyngeal carriage of *S. pneumoniae* among immunized children, which in turn decreases transmission to non-immunized persons [Rubin et al., 2010]. Availability of preventive procedure, however, is mainly restricted to developed countries, whereas the epidemic centers in Africa and India can only be covered punctually by special vaccination campaigns (WHO, 2011).

Increasing resistance against the first choice antibiotic penicillin complicates and raises costs of treatment of meningitis. Additional regimens with vancomycin plus a third-generation cephalosporin have become necessary [Bingen et al., 2008].

1.3 Pneumolysin - a bacterial protein toxin

1.3.1 Pore-forming toxins

Toxins are often defined as bacterial effectors that can act in the absence of the bacteria. Other bacterial effectors, in contrast, depend on delivery into the host, and thus on the presence of the pathogen (reviewed in [Aktories et al., 2011]).

Pore-forming protein toxins are an aggressive approach of pathogens, meant to enter, to damage/destroy, or to manipulate host cells for their own advantage. Those proteins encounter biological membranes and are known to lead to either membrane collapse, or to induction of signaling events downstream of membrane binding events, e. g. activation of small GTPases (reviewed in [Aktories and Barbieri, 2005]). Pore-forming proteins were identified in a wide range of organisms, including bacteria, plants, fungi, and even animals [Iacovache et al., 2008].

1.3.2 Pneumolysin - a member of the CDC family

The protein toxin pneumolysin (PLY) (471 amino acids, 53kD) represents one of the major virulence factors of *Streptococcus pneumoniae*. It belongs to the family of cholesterol-dependent cytolysins (CDCs), β-barrel pore forming toxins, secreted by Gram-positive bacteria via a - for *S. pneumoniae* - not yet verified mechanism [Jefferies et al., 2007; Marriott et al., 2008].

The toxins are released as water soluble monomers, that recognize cholesterol as a receptor on the target membrane, bind and oligomerize to characteristic arcs and ring-like structures (about 30-50 monomers), representing cell lysis inducing pores (with a diameter of about 260 Å) (reviewed in [Alouf, 2000; Giddings et al., 2003; Tweten, 2005]).

The pore forming event includes the insertion of two amphipathic β-hairpins from each of the ~35 monomers and spanning of the membrane via dramatic conformational changes in the protein. The events create a large transmembrane β-barrel that perforates the membrane [Tilley et al., 2005].

Pneumolysin is a four-domain protein. Its structure has not yet been solved in detail, but was fitted to the well investigated perfringolysin O (PFO) [Rossjohn et al., 2007], which is also a member of the CDCs. Pneumolysin is believed to be released via bacterial autolysis induced by the bacterial virulence factor autolysin (Lyt A), during the late log-phase of bacterial growth [Berry et al., 1989]. Also the treatment with cell wall degrading antibiotics, such as penicillin, can induce release of high amounts of the toxin [Mitchell and Mitchell, 2010]. Upon cholesterol-recognition (proper binding requires at least 30 mole % of membrane cholesterol [Alving et al., 1979]), monomers of PLY oligomerize on the host cell membrane.

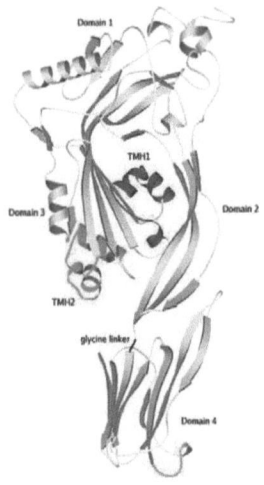

Fig. 4: Protein configuration of pneumolysin monomer.
Structure fitted to perfringolysin O. PLY consists of four domains. Domain 4 binds membrane cholesterol with high affinity. Domain 1 and 2 are responsible for oligomerization and domain 3 penetrates the membrane by unfolding of the transmembrane hairpin regions (TMH) 1 and 2.
(Source: [Rossjohn et al., 2007])

The membrane recognition motif in domain 4 (D4) is discussed controversially. It was widely believed that a highly conserved tryptophan-rich undecapeptide sequence at the base of D4 mediates cholesterol recognition and binding [Jacobs et al., 1999]. Recent studies, however, show that a threonin-leucin pair (at position 459/460 in PLY), that is conserved in all CDCs and located in loop 1 of D4, is responsible for membrane binding of the monomers [Farrand et al., 2010]. Oligomerization and penetration involves domains 1-3 (D123), with D3 dipping through the membrane and thus destroying membrane integrity.

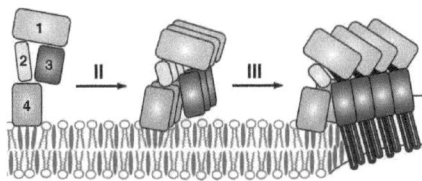

Fig. 5: Pore-formation mechanism.
Toxin monomers bind to cholesterol on the PM and oligomerize (II). This leads to a conformational change, which induces the penetration of the membrane by unfolding of domain 3 of each monomer (III). (Source: [Heuck et al., 2010], modified)

1.3.3 Role in pathogenicity

Pneumolysin plays an extraordinary role in the pathogenicity of *S. pneumoniae*. Lack of the toxin lowers induction of pro-inflammatory response and tissue inflammation in general, the production of reactive oxygen species and other immune mediators, and thus hampers a proper bacterial clearance. Consequently, PLY deficient strains are associated with improved survival, lower morbidity and decreased numbers of bacteria in the blood [Canvin et al., 1995; Kadioglu et al., 2000].

The toxin is, as well, the main trigger of immune responses against *S. pneumoniae*. Toll-like receptor 4 (TLR-4) and the TLR adaptor myeloid differentiation factor 88 are known to be essential for macrophage expression of Tumor Necrosis Factor-alpha (TNF-α) and Interleukin-6 (IL-6) in response to PLY [Malley et al., 2003], but also TLR-2 is supposed to play a role in bacterial clearance [van Rossum et al., 2005].

T-cells represent the main cellular response to pneumococcus, and T-cell depletion also reduces bacterial clearance from infection sites [van Rossum et al., 2005]. Furthermore, PLY is a major target for anti-pneumococcal antibodies. HIV-1 infection is accompanied by impaired production of anti-PLY antibodies, which explains higher susceptibility to invasive pneumococcal disease of HIV patients [Amdahl et al., 1995]. Pregnant women with high antibody levels against PLY are less likely to have infants who become colonized with *S. pneumoniae* during their infancy, and antibodies, in general, are thought to be important in the clearance of colonization with the bacterium [Dagan et al., 2002]. PLY is additionally a very potent activator (but also in-activator) of the classical complement pathway, as it shows sequence homology with the acute phase protein, C-reactive protein (CRP), and thus can bind to the F_c portion of immunoglobulin and to C1q [Paton et al., 1984]. This leads to an impaired opsonic activity of serum PMNLs (polymorphonuclear leukocytes) *in vivo*, which in turn has a strong influence on bacterial clearance [Paton et al., 1984].

Table 1: **Pathogenicity of S. pneumoniae.** *[Mitchell and Mitchell, 2010]*

Virulence factor	Effect
polysaccharide capsule	• protection from phagocytosis by prevention of binding of iC3b (complement system) and F_c (immunoglobulin) • enhancement of colonization • prevention of mechanical removal by mucus • restriction of autolysis
pneumolysin (PLY)	high, lytic concentrations: • pore formation and cell lysis/death low, sub-lytic concentrations: • activation of the classical complement pathway • promotion of inflammation • production of reactive oxygen intermediates (both mediated by interaction with TRL-4) • astrocytic process retraction, cortical astroglial reorganization, interstitial tissue retention ⇨ enhanced penetration of bacteria and toxic macromolecules [Hupp et al., 2012], and brain damage in general
hyaluronidase	• enhancement of spreading and colonization
neuraminidase A (NanA)	• enhancement of colonization

	• biofilm formation
pilus	• enhancement of binding to epithelial cells • stimulation of proinflammatory cytokines
lipoproteins (PsaA, SirA, PpmA)	• adhesion to cells • colonization
autolysin (LytA)	• release of PLY from the bacteria • release of inflammatory cell wall degradation products
pneumococcal surface proteins (PspA, PspC)	• interaction with the complement • adhesion
genome variation	• evolutionary advantages • resistance against antibiotics

1.4 The cytoskeleton

1.4.1 General properties

Proper functioning of a cell is dependent on its adaptability and motility, its mechanic stability, its ability to communicate with its environment and to accurately organize its interior. These properties are provided to the cell by the components of its cytoskeleton, which consists of three different filament systems, working collectively together.

The cytoskeleton tears the chromosomes and dividing cells apart during mitosis, it drives trafficking of organelles and molecules throughout the cytoplasm, supports the fragile plasma membrane and is part of the intercellular connections and mechanical linkages, that help the cell to communicate and bear mechanic stresses and tensions, that occur when the environment shifts and changes. The cytoskeleton enables cells to crawl across surfaces and sperms to swim. It drives axon and dendrite growth, and last, but not least, controls the diversity of eukaryotic cell shapes.

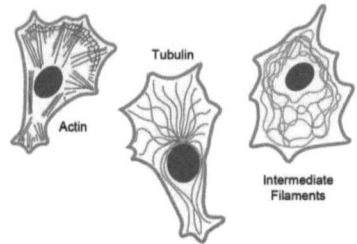

Fig. 6: Schematic of the components of the cell cytoskeleton.
Actin filaments, microtubules and intermediate filaments provide mammalian cells numerous essential skills, like motility, molecular trafficking and stability.
(Source: [Alberts et al., 2008], modified*)

The cytoskeleton is divided into three subgroups, intermediate filaments (e. g. GFAP in astrocytes), microtubules, and the microfilaments (actin) (Fig. 6). Intermediate filaments provide mechanical strength to the cell. They line the inner face of the nuclear envelope to form a protective cage for the DNA. Microtubules position intracellular organelles, and are responsible for trafficking, they build the mitotic spindle, and form motile whips called cilia and flagella. Microfilaments, fibers composed of actin, are mainly positioned below the plasma membrane (building the actin cortex), thus protecting the lipid bilayer from mechanic harm and enabling movement of the whole cell by formation of large actin aggregations that push the membrane forward. The filaments are flexible and can easily be bent. They can be organized into 2-dimensional networks or 3-dimensional gels.

If those dynamic aggregations are consisting of parallel actin bundles, they build membrane spikes, which are called filopodia. Mesh-like aggregations built from branched actin filaments are called lamellipodia. Long actin bundles, reaching from one to the other side of the cell comprise stressfibers, often established when tensions are affecting the cell, and under conditions of active movement.

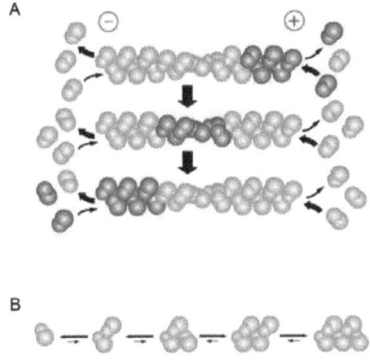
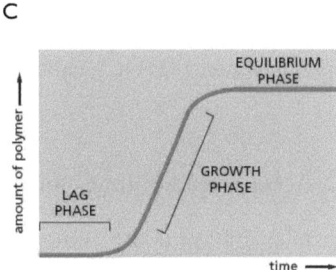

Fig. 7: Actin treadmilling and nucleation.
A. Treadmilling. *In the absence of actin effectors subunits undergo a net assembly at the plus end and a net disassembly at the minus end at an identical rate. The polymer maintains a constant length, even though there is a net flux of subunits through the polymers (dark red).*
B. Nucleation. *Two actin monomers bind relatively weakly to each other, but addition of a third monomer makes the group more stable. Further monomer addition leads to the formation of a stable "nucleus".*
C. Kinetics. *The formation of a nucleus is very slow (explains the lag phase seen during actin assembly). The lag phase is followed by rapid actin polymerization, resulting in an equilibrium phase. ([Alberts et al., 2008], modified*)*

As already mentioned, the cytoskeletal systems are no static structures, but very dynamic and adaptable. They can either persist for a longer time, or change spontaneously, according to need. Alberts et al. use a very suitable example, comparing the cytoskeleton to an ant trail, extending from an ant nest to a selected picnic site. Such trail can persist for many hours, but can also be rearranged spontaneously, if a better picnic site is discovered. Like the single molecules that compose the cytoskeletal structures, every single ant will be concerted into a new trail.

Intermediate filaments are ropelike structures of about 10 nm diameter, composed of intermediate filament proteins, a large and heterogeneous family of proteins.

Microtubules (MTs), which are hollow cylinders, are made of the protein tubulin, and have an outer diameter of 25 nm. One microtubule end is usually attached to a MT-organizing center (MTOC), called a *centrosome* [Alberts et al., 2008].

1.4.2 Actin - treadmilling - nucleation

Actin microfilaments are composed of actin protein subunits, aggregated into two-stranded helical polymers with a diameter of 5-9 nm. Of the three isoforms of actin, only the β- and the γ-form are part of cytoskeletal structures. The α-isoform is part of muscle tissue and the contractile apparatus. Actin subunits are globular polypeptide chains with an ATP (Adenosine triphosphate) nucleotide binding site. They assemble in a head-to-tail manner to generate filaments with a distinct structural polarity. Two actin protofilaments twist around each other in a right-handed helix to form one actin filament (Fig. 8). This determines two possible forms of actin, G-actin (monomeric) and F-actin (filamentous).

The orientation of the monomers within the chain gives the filament a structural polarity, where the kinetic rate constants for association and dissociation are often much greater at one end than at the other. The more dynamic end of a filament, where both growth and shrinkage are fast, is called the plus end (also barbed end), and the other end is called minus end (also pointed end).

The ATP-binding cleft on the actin monomer points to the minus end. Actin subunits have an enzymatic phosphatase activity, releasing phosphate groups from the ATPs. The hydrolysis proceeds very slowly for the monomers, but the process is accelerated as soon as a subunit is incorporated into a polymer. The free phosphate (P_i) is released, but the nucleotide diphosphate (ADP) remains trapped in the filament structure.

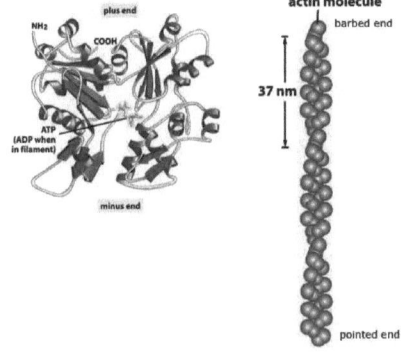

Fig. 8: Actin monomer and actin filament.
In the actin monomer ATP/ADP is bound in a deep cleft in the center. Each monomer has a certain orientation. The arrangement of monomers leads to the formation of protofilaments. Two of those wind around each other to form an actin microfilament, with a twist repeating every 37 nm. (Source: [Alberts et al., 2008], modified)*

Thus two different types of filament structures exist, one with ATP bound and one with ADP bound. When a nucleotide is hydrolyzed, much of the free energy is stored in the polymer lattice.

If the rate for subunit addition is high (that is if the filament is rapidly growing), then it is likely that a new subunit will add on to the polymer before the nucleotide in the previously added subunit has been hydrolyzed, so that the tip of the polymer remains in the ATP-bound form (ATP cap).

The rate of subunit addition at the end of a filament is the product of the free subunit concentration and the rate constant k_{on}. The k_{on} is much faster for the plus end of a filament than for the minus end because of a structural difference between the two ends. The ADP-bound form leans more readily toward disassembly, while the ATP-bound form leans more readily toward assembly. If the concentration of free subunits in solution is in an intermediate range, the filament adds subunits at the plus end, and simultaneously loses subunits from the minus end. This leads to the phenomenon of *treadmilling*. Under conditions of "steady-state *treadmilling*", when addition and dissociation are in balance, the total length of the filament remains unchanged.

During *treadmilling*, subunits are recruited at the plus end of the polymer in the ATP-bound form and shed from the minus end in the ADP-bound form.

The ATP hydrolysis that occurs along the way gives rise to the difference in the free energy of the association/dissociation reaction at the plus and minus ends of the actin filament and thereby makes *treadmilling* possible.

The binding of two actin molecules is a relatively weak connection. But as soon as a third actin monomer is added to form a trimer, the entire group becomes more stable. Further monomer addition can take place onto this trimer, which then acts as a nucleus for polymerization. This phenomenon is called nucleation. The assembly of a nucleus is relatively slow, which explains the lag phase seen during polymerization of actin. During the growth phase monomers are added to the exposed ends of the growing filament, causing elongation. The equilibrium phase (steady state) is reached when the growth of the polymer precisely balances the shrinkage of the polymer (Fig. 7 C).

The rapid interconversion between a growing and shrinking stat, at a uniform subunit concentration, is called *dynamic instability*. The change from growth to rapid shrinkage is called a *catastrophe*, while the change to growth is called a *rescue*. Both *dynamic instability* and *treadmilling* allow a cell to maintain the same overall filament content, while individual subunits constantly recycle between the filaments and the cytosol. Actin filament turnover is a typically rapid process, with filaments persisting for only few tens of seconds or minutes. The single subunits themselves are small and can diffuse very rapidly. They can cross the diameter of a typical eukaryotic cell in several seconds [Alberts et al., 2008].

1.4.3 The actin cortex and active cell movement

The actin cortex is located directly below the plasma membrane of animal cells and comprises an aggregation of actin and actin binding proteins (ABPs) in a high density. It enables the cell to displace actively. Actin polymerization at the leading edge forms a lamellipodium, and its firm attachment to the substratum moves the edge forward and stretches the actin cortex. Contraction at the rear of the cell pushes the body of the cell forward to relax some of the tension. New focal points develop at the front and old ones are disassembled at the rear. Repetition of the cycle moves the cell forward in a stepwise fashion. Tight coordination of all steps leads to a smooth movement [Alberts et al., 2008].

As a place of massive actin accumulation the actin cortex also harbors a high number of actin regulator proteins, such as the Arp2/3 complex and Gelsolin (see below).

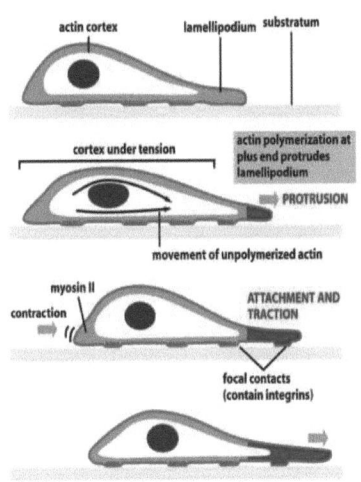

Fig. 9: Active movement of animal cells mediated by the actin cortex.

The actin cortex (orange) comprises the massive actin aggregation right below the PM. Protrusion of a cell is dependent on actin-polymerization at the leading edge of the cell, resulting in the formation of a lamellipodium, which firmly attaches to the extra cellular matrix. Contraction at the rear of the cell pushes the cell body forward to relax some of the tension (traction). At the front new focal adhesions (FAs) are established, and old ones disassembled at the back. Tight coordination of those steps move a cell forward in a smooth manner. The newly established cortical actin is shown in red. (Source: [Alberts et al., 2008]*)

1.4.4 Actin binding proteins (ABPs)

In vivo actin filaments, microtubules and also intermediate filaments are much more dynamic than *in vitro*. Reason for this is that for each filament system of the cytoskeleton exists a huge set of proteins controlling their turnover and structure. These so-called accessory proteins bind either to the filaments or their free subunits.

Overall, nucleation of monomers is the major step of cytoskeletal filament establishment. Actin nucleation occurs mainly at or near the plasma membrane, where the actin cortex is located. From this point also surface protrusions, like filopodia and lamellipodia, emerge (indicated in Fig. 12). Although nucleation is the initializing and most important step, there are many other cytoskeleton-related reactions to be taken care of, when external signals, that arise from changes in the cell environment, hit the cell surface and have to be processed inside the cell to remodel the cytoskeleton and trigger a reaction of the cell [Alberts et al., 2008]. In the following paragraph the most common of the numerous actin binding proteins will be characterized.

1.4.4.1 The actin-related protein complex (Arp2/3)

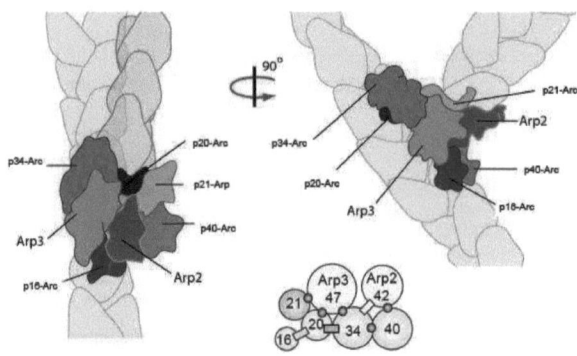

Fig. 10: *The Arp2/3 complex.*
The actin related proteins build a heptameric complex that assembles to potently nucleate actin polymerization, both de novo, *and originating from pre-formed actin filaments at a 70° angle, which leads to an actin meshwork, mostly seen at the leading edge of protruding cells.*
(Source: Manual of Cellular and Molecular Function, *The National University of Singapore, 2009; Illustrators: Pham Thi Phuong Thao & Theng Tze Yin Meryl (modified); 2011)*

Nucleation of actin can be catalyzed by two different APBs, the formins and the ARP (actin-related protein) complex. The Arp complex (also known as Arp2/3 complex) is a seven-protein complex, including two actin-related proteins, the ARPs (2 and 3), each of which is about 45% identical to actin. The five additional subunits of the complex are p41-Arc, p34-Arc, p21-Arc, p20-Arc and p16-Arc.

The ARPs nucleate actin filament growth from the minus end by capping this side, allowing rapid elongation at the other side. Arp can also attach to the side of another actin filament while remaining bound to the minus end of the filament that it has nucleated, thereby establishing a branched web. The ARP complex is localized in regions of rapid actin growth, such as filopodia and lamellipodia. Signaling molecules and effectors at the cytosolic face of the plasma membrane regulate its activity. *In vivo* and especially *in vitro* the Arp complex requires activation by binding of the conserved C-terminal verprolin-homology-cofilin-homology-acidic (VCA) domain of an N-WASP (Neuronal Wiskott-Aldrich-Syndrome protein) family protein to one subunit of the Arp complex [Alberts et al., 2008; Condeelis, 2001; Mullins et al., 1998]. N-WASP possesses several domains, interacting with different factors: a pleckstrin homology (PH) domain that binds phosphatidylinositol(4,5)bisphosphate (PIP_2), a Cdc42-binding domain (GDB) domain, a proline-rich region, a G-actin-binding verprolin homology (V) domain, a domain (C) with homology to the actin-depolymerizing protein cofilin, and a C-terminal acidic segment (A) [Rohatgi et al., 1999].

1.4.4.2 Gelsolin

Actin filament severing (a process contrary to nucleation) is processed by the gelsolin superfamily of actin binding proteins, whose severing activity is activated by high levels of cytosolic Ca^{2+} [Alberts et al., 2008]. It is one of the most potent actin filament severing proteins known. Gelsolin binds to one side of an actin filament, and severs the filament after kinking [McGough et al., 1998]. After the severing it stays attached to the barbed end as a cap. This results in short filaments, which can neither re-anneal nor elongate their plus ends. This in turn leads to a disassembly of the microfilament system. However, gelsolin can also have a filament constructive effect, as it increases the number of filaments. In case of an uncapping event, the high number of previously generated short filaments, comprises a high number of new nucleation sites or polymerization-competent ends, from which actin can grow and rebuild the microfilament system [Sun et al., 1999]. It is known that gelsolin is a downstream effector of the small GTPase Rac, with calcium ions and phosphoinositides as intermediate regulators [Hartwig et al., 1995].

1.5 How bacterial toxins intervene with cellular functions

1.5.1 Toxin strategies

As mentioned above, bacterial effectors have evolved to alter cell function for their own advantage. In this context bacterial toxins have achieved special importance, as being released as soluble protein monomers or compounds, they have free access to cell membranes, receptors, cell signaling, or even intracellular molecules or structures, in case they are translocated to the cytosol via, for example, a type-III secretion system (TTSS) or endocytosis. Generally spoken, bacterial effectors can influence the fate of their host cells, favoring either death or survival, depending on their needs [Fiorentini et al., 2003].

Mechanisms that lead to the collapse of cells are either induction of necrosis/apoptosis or direct and fast destruction of membranous structures and thus initiation of cell lysis. Latter is the major approach of pore-forming toxins, like the CDCs (introduced in paragraph 1.3). Via formation of big ring-like structures on the membrane and penetration through the lipid-bilayer they form holes that are big enough (~ 260 Å, 26 nm) to kill the cell. Other toxins follow different strategies to alter cell function without cellular destruction. A major target in this context is the cell cytoskeleton, especially the actin cytoskeleton, which plays a pivotal role in cell stability, cell motility, in intracellular vesicle transport, phagocytosis, and mitosis (see paragraph 1.4). Toxins approach the actin scaffold either directly, or they hijack its endogenous effector proteins, as described below.

Invasive toxins of *Salmonella* (SipA and SipC) interact directly with actin molecules, leading to filament bundling and nucleation (SipC), or in case of SipA to a polymerization and stabilization of F-actin filaments via reduction of the critical concentration for filament formation. Furthermore, SipA functions as a potentiator of SipC, leading to stronger nucleation and F-actin bundling [Hayward and Koronakis, 1999; Lilic et al., 2003; McGhie et al., 2001].

ADP-ribosylation of actin, which is, for example, carried out by the *Clostridium botulinum* C2 toxin and other binary toxins, leads to depolymerization of actin and inhibits its polymerization. G-actin, which is ADP-ribosylated at Arg177, cannot polymerize, but it can still bind to and cap the barbed ends of pre-existing actin filaments. Thus, further growth of filaments from these ends is inhibited, leading to destruction of the actin cytoskeleton. This event in turn leads to release of growing microtubules at the cell membrane. Thus, long protrusions extending from the cell membrane, develop, which in turn facilitate adherence and colonization of the bacteria [Aktories et al., 1986; Schwan et al., 2009].

A third mechanism that leads to direct actin modification, namely actin nucleation, is carried out by the Tir toxin of *Escherichia coli*, ActA of *Listeria* spp., and IcsA of *Shigella flexneri*. In this form of actin nucleation, however, a third component, actin effector proteins like α-actinin or Arp2/3 are involved. Function of the Arp complex is explained in chapter 1.4.4.1 of this work. It is known that actin pedestal formation of enteropathogenic *E. coli* is mediated via recruitment and activation of Arp complex and WASP (Wiskott-Aldrich syndrome proteins) by the Tir toxin [Kalman et al., 1999].

Bacterial effectors that do not directly interact with actin, act on its regulators, the so called Rho proteins or Rho family of small GTPases.

1.5.2 Small GTPases

Small GTPases (20 to 30kD GTP-binding proteins), which comprise eukaryotic molecular switches, have become a major target of toxins, as they regulate substantial events, such as cell migration and chemotaxis, cell division, neurite and axon outgrowth, regulation of cell polarity, phagocytosis and endosome trafficking, but also cancer progression and metastasis (reviewed in [Heasman and Ridley, 2008]).

Approximately 1% of the human genome encodes proteins that either regulate or are regulated by direct interaction with members of the Rho family GTPases, which is a distinct family within the superfamily of Ras-related small GTPases. Members of the Rho (Ras homology) family of small GTPases (which comprise at least 20 distinct proteins) function as regulators of the cell cytoskeleton (for further review see [Jaffe and Hall, 2005]). Best-characterized in this group are Rho, Rac and Cdc42.

Rho was shown to be responsible for the assembly of contractile actin and myosin filaments (stress fibers) [Ridley and Hall, 1992], whereas Rac induces cell surface protrusions at the leading edge of a cell via enrichment of actin structures below the protruding membrane (lamellipodia) [Ridley et al., 1992].

Cdc42 is known to induce the formation of finger-like actin-membrane structures, which are called filopodia [Nobes and Hall, 1995]. The Rho switch is very tightly regulated. The human genome contains more than 60 activators (GEFs: guanine nucleotide exchange factors) and more than 70 inactivators (GAPs: GTPase activating proteins). Over 60 proteins were identified to be targets of the Rho family GTPases (reviewed in [Etienne-Manneville and Hall, 2002]).

The control mechanism by which those molecular switches function to regulate complex cellular signaling events is based on one of the most abundant biochemical strategies, namely the cleavage of energetic phosphate-bonds. During the GTP (guanosine tri-phosphate) bound state, the GTPases recognize their target molecules and bind those (on-state), until the hydrolysis of GTP to GDP (guanosine di-phosphate) turns them to the off-state and leads to the release of target proteins. In this context GEFs promote exchange of nucleotides and thus activation, whereas GAPs stimulate the formation into the inactive form via hydrolysis of nucleotides

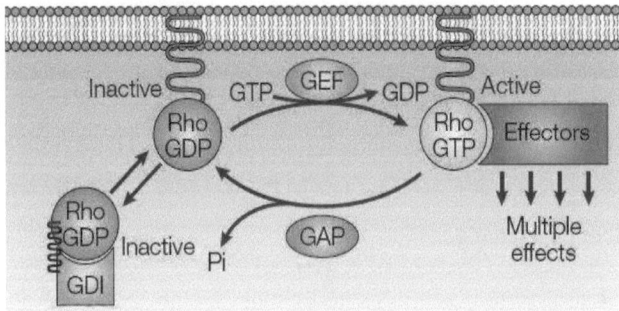

Fig. 11: *Molecular mechanism of small GTPase activation and deactivation.*
The transition of small GTPases ("Rho") from an inactive into an active form is mediated by GEFs (guanine nucleotide exchange factors) and GAPs (GTPase activating proteins). GEFs activate GTPases by promoting the release of GDP and the binding of GTP, which in turn induces interaction with their downstream effectors. GAPs inactivate Rho GTPases by increasing the intrinsic GTPase activity of Rho proteins. Guanine nucleotide-dissociation inhibitors (GDIs) bind to Rho proteins, sequestering them in the cytoplasm away from their regulators and targets. (Source:[Aktories and Barbieri, 2005])

A third group of GTPase effector molecules are the GDIs (guanine nucleotide dissociation inhibitors), of which exist at least four different. GDIs bind inactivated GTPases in the cytosol, and consequently prevent translocation to the membrane and thus activation.

The switch function is based on an interposition of the Rho protein between upstream and downstream effectors or signals. An upstream signal stimulates the dissociation of GDP from the inactive form, which is followed by the binding of GTP, leading to a conformational change of the downstream effector-binding region, so that this region gets accessible to downstream effector molecules [Takai et al., 2001].

1.5.3 Toxin effects on small GTPases

As small GTPases are of major importance for cell behavior, it is obvious that bacteria have evolved mechanisms to intervene with proper function of those. Hijacking of small GTPases by bacterial toxins is carried out on two different modes of action. Bacterial effectors either directly mimic small GTPase molecules, or act as their activators or inactivators (GEFs or GAPs), leading to disturbance of normally regulated nucleotide exchange.

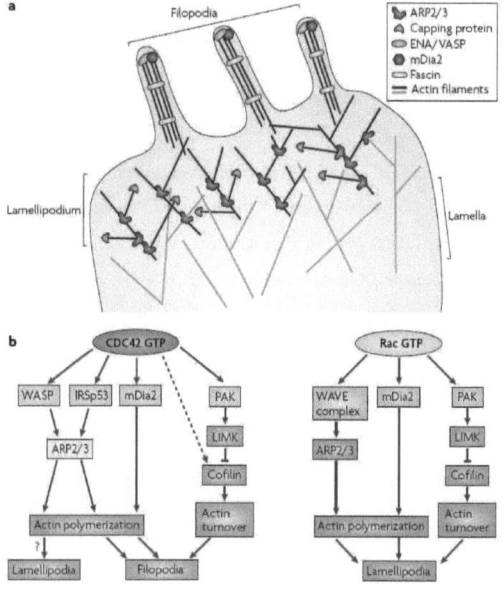

Fig. 12: Mechanism of actin modification by Rac and Cdc42.

a. Actin structures built up at the leading edge of the cell. Within the lamellipodia the Arp complex and other actin effectors lead to the formation of an actin mesh (branched actin structures). Out of the lamellipodia emanate so called filopodia, which contain parallel bundles of actin.

b. Upon activation by their effector proteins (GEFs), small GTPases activate downstream effector proteins like WASP and Arp, or, e. g., proteins of the kinase family like the LIM kinase. Those proteins in turn influence the actin turnover and lead to actin cytoskeleton remodeling. (Source: [Heasman and Ridley, 2008], modified)

They also carry out ADP-ribosylation, glucosylation, proteolysis, adenylation, deamidation, or tansglutamination reactions on Rho proteins directly [Aktories et al., 1989; Just et al., 1995; Masuda et al., 2000; Schmidt et al., 1997; Shao et al., 2002; Yarbrough et al., 2009]. Approaching the switch molecules in such dramatic manner allows hijacking and alteration of cellular functions.

Table 2: **Bacterial toxins targeting Rho GTPases.** *[Aktories and Barbieri, 2005], modified*

Toxin	Mode of action	Substrates
Clostridial glucosylating toxins		
Toxin A and B (*Clostridium difficile*, strain VPI 10463), Haemorrhagic toxin (*Clostridium sordellii*)	Glucosylation of Rho GTPases (for example, RhoA at Thr37)	RhoA, Rac, Cdc42
Toxin B 1470 (*C. difficile* strain 1470)	Glucosylation of Rho/Ras GTPases (for example, Rac at Thr35)	Rac, (Cdc42), Ha-Ras, R-Ras, Ral, Rap but not RhoA
Lethal toxin (*C. sordellii*)	Glucosylation of Rho/Ras GTPases (for example, Rac at Thr35)	Rac, (Cdc42), Ras, Ral, Rap but not RhoA
α-Toxin (*Clostridium novyi*)	N-acetylglucosaminylation of Rho GTPases (for example, RhoA at Thr37)	RhoA, Rac, Cdc42
Rho-ADP-ribosylating toxins		
C3 exoenzymes: C3bot (*Clostridium botulinum*), C3lim (*Clostridium limosum*), C3cer (*Bacillus cereus*), C3stau (*Staphylococcus aureus*)	ADP-ribosylation of Rho at Asn41	RhoA, RhoB and RhoC C3stau also RhoE/Rnd3
Rho-targeting protease		
YopT (*Yersinia enterocolitica*)	Cleavage of the C-terminal isoprenylated Cys	Rho, Rac, Cdc42
Rho-deamidating/transglutaminating toxins		
CNF1, CNF2 (*Escherichia coli*) CNFY (*Yersinia pseudotuberculosis*)	Activation of Rho GTPases by deamidation (for example, RhoA at Gln63)	Rho, Rac, Cdc42; RhoA in 12: intact cells
DNT (*Bordetella* spp.)	Activation of Rho GTPases by transglutamination (for example, RhoA at Gln63)	Rho, Rac and Cdc42
Toxins modulating the nucleotide state of Rho GTPases		
SopE, SopE2 (*Salmonella*)	GEFs for Rho GTPases	Rho, Rac, Cdc42
ExoS, ExoT (*Pseudomonas aeruginosa*)	GAPs for Rho GTPases, additional ADP-ribosyltransferase activity	ExoS, ExoT: GAPs for Rho, Rac, Cdc42; ExoS: ADP-ribosylation of Ras and ERMs; ExoT: ADP-ribosylation of Crk proteins
SptP (*Salmonella*)	GAPs for Rho GTPases. Additional tyrosine phosphatase activity	GAP for Rac and Cdc42; dephosphorylation of Erk and vimentin
YopE (*Yersinia* spp.)	GAP for Rho GTPases	Rho, Rac, Cdc42

Table 3: **Bacterial toxins targeting actin.** *[Aktories and Barbieri, 2005], modified*

Actin-ADP-ribosylating toxins		
C2 toxin (*C. botulinum*)	ADP-ribosylation of actin at Arg177	Non-muscle β/γ-actin
Iota toxin (*Clostridium perfringens*), spiroforme toxin (*Clostridium spiroforme*), VIP (*B. cereus*), CDT (*C. difficile*), SpvB (*Salmonella*)	ADP-ribosylation of actin at Arg177	Non-muscle and muscle α-actins
Crosslinking of actin		
RTX (*Vibrio cholerae*)	Formation of actin dimers and trimers by an unknown mechanism	Actin
Non-covalent modification of actin		
SipA, SipC (*Salmonella*)	Decrease in the critical concentration for actin polymerization (SipA); nucleation and bundling of actin (SipC)	Actin
Modulation of actin polymerization		
ActA (*Listeria* spp.)	Actin nucleation by direct activation of ARP2/3 complex	ARP2/3 complex
RickA (*Rickettsia conorii*)	Actin nucleation by direct activation of ARP2/3 complex	ARP2/3 complex
IcsA (*Shigella flexneri*)	Actin nucleation through ARP2/3 by recruiting N-WASP	N-WASP
A36R (vaccinia virus)	Actin nucleation through ARP2/3 by Nck and N-WASP recruitment	Nck

1.6 The brain

1.6.1 Brain anatomy

The brain is one of the sterile compartments of the body. Being the center of the nervous system and thus the control center of the body, protection demands special effort. Whereas the skull protects from mechanic injury, the blood-brain barrier (see below) comprises a barrier for all agents that potentially lead to disturbance of brain homeostasis. Further protection of the brain and whole nervous system is provided by the meninges, which are a set of membranes (Dura mater, Arachnoid mater, Pia mater), enclosing the components of the CNS. The subarachnoid space comprises the space between Arachnoid and Pia mater, and is filled with cerebrospinal fluid (CSF, liquor), which additionally provides a basic mechanical (cushion) and immunological protection to the brain. It occupies also the ventricular system around and inside the brain and spinal cord.

Macroscopically the brain consists of the two symmetric (right and left) hemispheres of the cerebrum (forebrain), and the cerebellum (Latin for little brain). The cerebrum is further divided into frontal, occipital, parietal, and temporal lobe, each of which harbors different neuro-physiological functions. The brain stem is the lower extension that connects with the spinal cord. The Medulla oblongata comprises the lower half of the brain stem. The thalamus is located between the cerebral cortex and the midbrain, whereas the Hypothalamus lies below the thalamus and synthesizes and secretes several neuro-hormones.

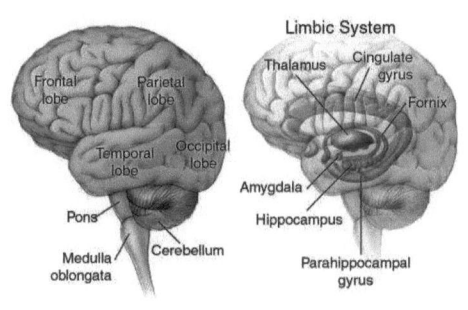

Fig. 13:   Anatomy of the brain.
The human brain is divided into the right and left hemisphere, and the cerebellum. The hemispheres are further divided into several lobes, each of which has unique physiological functions. The hippocampus is another paired structure and lies in the center of the brain within the Limbic System. It is, amongst others, responsible for memory and learning.
(Source: www.clas.ufl.edu/users/nnh /sem05/brnpic2-5.htm; 2011)

The cerebral cortex is a sheet of neurons comprising the outermost part of the cerebrum. It consists of up to six horizontal layers, each of which has a different composition in terms of neurons and connectivity. The cortex plays a key role in memory, attention, language and consciousness. The neurons in this sheet are unmyelinated, and thus appear to be gray ("gray matter"). The "white matter" below is formed by myelinated axons, which interconnect neurons in different regions of the cortex with each other and with neurons in other parts of the CNS.

Closely associated to the cortex lies the hippocampus (Fig. 13), which is also a paired structure with halves in the left and right sides of the brain. It contains two interlocking parts, the Ammon's horn and the dentate gyrus. The hippocampus is responsible for behavioral inhibition, attention, spatial memory, learning and navigation [Kandel, 1996; Squire, 2003].

1.6.2 Neuroglia

The human brain consists of about 10 billion neurons, however, not being the only component of the central nervous system (CNS). The term neuroglia or "nerveglue" was set by Rudolf Virchow in 1859, describing the "inactive connective tissue" holding neurons together in the CNS. Ramón y Cajal and del Rio-Hortega were later able to distinguish the different supporting cell types present in the tissue of the CNS: oligodentrocytes, astrocytes, and microglia. In the peripheral nervous system (PNS) the Schwann Cells are the major neuroglial component. Virtually all axons are ensheathed or fenced in by other cells types, providing support or trophic supply.

Astrocytes -which are described in more detail below- comprise the most abundant cell population in the brain. They interconnect with each other via gap junctions and with neurons and blood vessels through their endfeet, and are responsible for the maintenance of the blood-brain barrier and brain homeostasis [Sofroniew and Vinters, 2010; Squire, 2003].

Oligodendrocytes form with a single sheet of their plasma membrane the myelin sheaths around neurons, increasing the rapidity of conduction of nerve impulses along axons by many-fold.

Microglia are mediators of immune responses in nerve tissue, and therefore are often designated the macrophages of the brain and PNS. Microglia are resident in the healthy brain in a ramified phenotype, but can rapidly transform into migratory, ameboid and phagocytic forms by receiving an activation signal. They become activated in any kind of neurodegenerative condition. In the mouse brain microglia constitute 5-20% of the total cells [Squire, 2003].

1.6.3 Astroglia

1.6.3.1 Astrocytic features

One major group of supportive cells in the brain are astroglia, also called astrocytes. Astrocytes are star shaped, process bearing cells, that constitute from 20 to 50% of the volume of most brain areas, outnumbering neurons by a five-fold. There are no regions devoid of astrocytes in the CNS, but they virtually tile the nervous system in a non-overlapping manner.

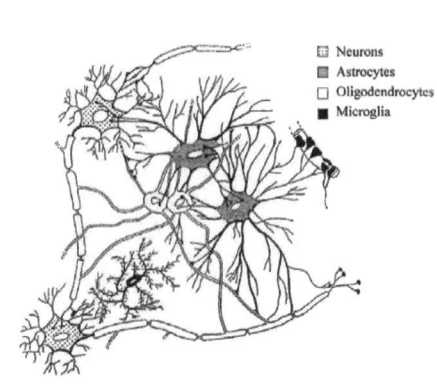

Fig. 14: Astroglia. Schematic of astrocytes within the neuronal network. The schematic shows the embedding of astrotcytes (green) in brain tissue. The cells are closely interconnected with neurons (dots), and astrocyte-end feet are tightly applied to the walls of capillaries. Microglia (black) comprise highly motile immune cells, and oligodentro-cytes (white) ensheath neurons with their PM (Source:[Baumann and Pham-Dinh, 2001], modified)

Although astrocytes can adopt about 11 different phenotypes, the protoplasmic and the fibrous forms are the most prominent. They develop from radial glial cells and serve as a scaffold for the migration and guidance of neurons during early development of the latter. In the mature brain, astrocytes fence in neurons and oligodentrocytes. The single astoglia cells are connected to each other via gap junctions, thereby forming a big syncytium throughout the brain, allowing ions and small molecules to diffuse across the brain parenchyma.

There are different distinct attributes that make it easy to identify astrocytes:
- their star-shaped exterior
- their end feet on capillaries
- GFAP: a special sort of intermediate filaments, only present in astrocytes
- S-100: a calcium binding protein

- a glutamine sythethase
- gap junctions consisting of connexin proteins
- desmosomes
- glycogen granules
- membrane orthogonal arrays
- the Aldh protein [Barres, 2008]

[Sofroniew and Vinters, 2010; Squire, 2003]

1.6.3.2 Astrocyte - neuron interaction

As already mentioned, there is an intimate relation between neurons and astrocytes, in which the neurons are highly dependent on the release of growth factors and re-uptake of neurotransmitters by the astrocytes. During neurotransmission, neurotransmitters and ions are released into the synaptic cleft in high amounts. Fast and proper cleanup of those substances afterwards are of high importance for the recycling and avoidance of interference with upcoming synaptic events. One of the major tasks of astrocytes in this context is the re-uptake of glutamate (besides e. g. GABA and ATP). They convert glutamate into glutamine and afterwards pour it into the extracellular space, where it is taken up by neurons. Neurons in turn use the glutamine to generate glutamate and γ-aminobutyric acid (GABA), which comprise potent excitatory and inhibitory neurotransmitters, respectively. This recycling process ensures proper excitability of neurons.

Uptake of sodium through ion channels and removal via gap junctions over the syncytium leads to a so called "spatial buffering". Additionally to the K^+-channels, astrocytes contain channels for Na^{2+}, Cl^-, HCO^{3+}, and Ca^{2+}. In response to stimuli, calcium waves are generated in astrocytes, thought to be mediated by secondary messengers, diffusing along gap junctions. Those kind of inter-cellular connections have also been detected between astrocytes and neurons, thus being not only participating in astrocyte-astrocyte, but also in astrocyte-neuron physiology.

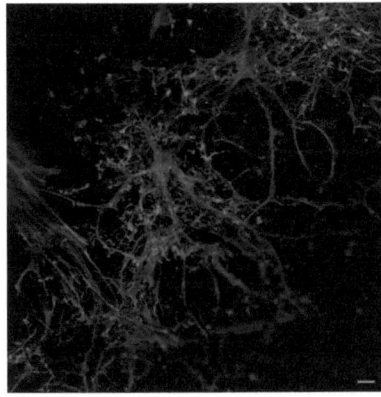

Fig. 15:   Co-localization of neurons and astrocytes.

Astrocytes and neurons are tightly associated in brain tissue. Astrocytes support neurons mechanically and metabo-trophically. Dysfunction of astrocytes will result in neuron instability. Scale bar: 15 µm.

(neurons: anti-actin/Cy3, red; astrocytes: anti-GFAP/FITC, green)

Besides the metabo-trophic and structuring functions, astrocytes are known to be integral functional elements of the synapses. They respond to neuronal activity and regulate synaptic transmission and plasticity. There is a bi-directional communication happening between the two cell types, leading to an active involvement of astrocytes in processing, transfer, and storage of information by the nervous system (reviewed in [Araque and Navarrete, 2010; Squire, 2003].

1.6.3.3 Astrogliosis

In case of neuronal injury, astrocytes take over an active part in wound closure by formation of a glial scar. Reactive astrogliosis is a hallmark of diseased CNS tissue, and is defined as an abnormal increase in the number of astrocytes due to the destruction of nearby neurons. It occurs in response to all kind of forms and severities of CNS damage and disease, even to minimal perturbations, and leads to molecular, cellular and functional changes in astrocytes. Depending on the form of insult, the reactive astrocytes undergo changes that range from alteration in molecular expression, over hypertrophy to enhanced proliferation and scar formation. Astrogliosis is regulated in a context-specific manner by extra- and intracellular signaling molecules (reviewed in [Sofroniew, 2009]).

1.6.3.4 The blood-brain barrier

The brain comprises a sterile, well-protected compartment with highly limited access for all kinds of agents that could imbalance brain homeostasis. This limitation is guaranteed by the blood-brain barrier (BBB), which is formed by brain epithelial cells lining the cerebral microvasculature. It protects the brain from fluctuations in plasma composition, which could lead to neuronal dysfunction.

Complex tight junctions between the adjacent endothelial cells force the major part of the molecular traffic to take a transcellular route across the BBB, rather than moving paracellularly through the junctions, as it happens in most other endothelia. Small molecules as oxygen, carbon dioxide or ethanol and lipophilic drugs can diffuse freely through the lipid membrane.

The traffic of hydrophilic molecules, in contrast, is regulated by specific transport systems on the luminal and abluminal membranes. Astrocytic perivascular end feet are of particular importance for

BBB stability. Astrocytes can upregulate many features of the barrier, leading to tighter tight junctions (physical barrier), the expression and polarized localization of transporters (transport barrier), and specialized enzyme systems (metabolic barrier). Their end feet are closely applied to the microvessel wall and show characteristic features for that location, including a high density of orthogonal arrays particles (OAPs) containing the water channel aquaporin 4 (AQP4) and the Kir4.1 potassium channel, which are involved in ion and volume regulation (reviewed in [Abbott, 2002; Abbott et al., 2006]).

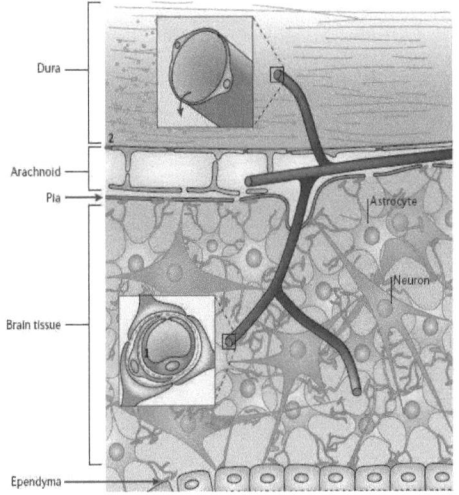

Fig. 16: ***Barrier sites in the CNS.***
(1) BBB: The brain epithelium forms the blood-brain barrier, which is strongly supported by astrocytes.
It comprises a highly selective barrier, protecting the brain from the entry of large hydrophilic molecules such as proteins and peptides, hydrophilic drugs and especially from the penetration of pathogenic agents.
(2) The arachnoid epithelium forms the middle layer of the meninges [Abbott et al., 2006].
(modified version of [Segal, 1990])

1.7 Basics of this study

Learning and memory are highly dependent on establishment and maintenance of synaptic contacts, and the most complex form of these synaptic contacts are located in the cortex. In this region the synaptic communication density and complexity is highest. Being positioned in an intimate proximity to the meninges and the subarachnoid space, where the pneumococci proliferate in the course of bacterial meningitis, and thus being exposed to substantial amounts of pneumolysin, the question arises whether the action of the toxin results in synaptic changes. Preliminary evidence demonstrates the existence of dendritic spine changes upon pneumolysin challenge, meaning it has a direct toxic effect on neuronal cells [Gerber and Nau, 2010; Iliev et al., 2009]. Furthermore, pneumolysin is known to induce cytotoxicity to microvascular epithelial cells of the BBB, which allows the bacteria to easily gain access to the brain [Zysk et al., 2001]. Astrocytes, as a part of the BBB, and as an essential brain sub-population, therefore comprise a main subject of interest when investigating bacterial meningitis. It is now known that the reshaping of astrocytes by PLY induces the remodeling of the structure of host brain tissue and induces interstitial fluid retention. Those alterations improve the penetration of toxic macromolecules and bacteria into the brain tissue [Hupp et al., 2012].

*: "Copyright 2008 from Molecular Biology of the Cell by Alberts et al. Reproduced by permission of Garland Science/Taylor & Francis Books, LLC."

2 METHODS

1.8 Pneumolysin preparation

Pneumolysin (PLY) and all variants of the toxin were friendly provided by Prof. Timothy Mitchell (University of Glasgow) as purified proteins in phosphate buffered saline (PBS). PLY was prepared as follows. The toxin was expressed in *E.coli* XL-1 cells and purified by hydrophobic interaction chromatography as described in [Mitchell et al., 1989]. The LPS content of purified toxin was determined by using the *Limulus* Amebocyte Lysate Kinetic-QCL Kit. All purified proteins had less than 0.6 endotoxin units/µg of protein. The purified wild-type toxin had an activity of about 5×10^4 hemolytic units/mg. One hemolytic unit was defined as the minimum amount of toxin needed to lyse 90% of 1.5×10^8 human erythrocytes per ml within 1 h at 37°C. The non-toxic delta6 version of PLY was generated by site-directed mutagenesis. Truncated mutants D1-3 (D123; a toxin fragment containing domain 1, 2 and 3), D4 (the domain 4 fragment of PLY), and GFP-tagged variants were purified similarly.

1.9 Pneumolysin concentration and lytic capacity determination

Toxin charges in general differ in their cytotoxicity, the capacity to lyse cells, which is dependent on the presence of active toxin molecules within the charge. In this work two different toxin charges were used, each of which had a different stock concentration and lytic capacity, as judged by LDH (lactate dehydrogenase) tests. With the help of propidium iodide (PI), the lytic capacity of different PLY concentrations was determined. Propidium iodide binds to DNA by intercalating between the bases with little or no sequence preference and with a stoichiometry of one dye per 4–5 base pairs of DNA. PI is membrane impermeant and generally excluded from viable cells. PI is commonly used for identifying dead cells in a population and as a counterstain in multicolor fluorescent techniques. With a size of 668 kDa, it has the capacity to cross the membrane of the cells via the pores formed by PLY.

PI was applied to mock-treated or PLY-treated cells in a dilution of 1 µl per 1 ml, directly into the imaged chamber. Subsequently the number of permeabilized cells was determined with the help of the ImageJ software. The total number of cells was determined by Hoechst nuclei staining. In order to define sub-lytic concentrations of the toxin, amounts between 0.05 and 0.5 µg/ml were applied to the cells and permeabilization was determined. For the first toxin charge, 0.1 µg/ml of PLY were assigned to be sub-lytic. As the second PLY charge was found to contain less active toxin molecules, the concentration had to be adapted in order to be equivalently toxic as the first charge. Thus, 0.2 µg/ml of PLY were used for the following experiments.

1.10 Labeling of PLY with Atto488

PLY-Atto488 was labeled and provided by Christina Förtsch, University of Würzburg.

Atto488 as NHS-Ester was added to PLY in a 4-fold molar excess (in 0.1 M bicine buffer, pH 9). Alkaline conditions were maintained during the labeling reaction for 30 min at room temperature in the dark. The labeling reaction was terminated by adding Tris-Cl (final concentration 10 mM, pH 8.0). The excess dye was removed by using a gel-filtration column. Covalent bonding was verified by an SDS-PAGE of the eluted fractions and examination on an UV-transilluminator. Concentration of labeled toxin was calculated after protein determination with BCA^{TM} Protein Assay Kit. Detailed analysis of membrane binding, lytic capacity and labeling confirmed that the labeled toxin behaved identically with the wild-type PLY.

1.11 Preparation of cells

All preparation steps were carried out under a binocular with an external light source, and in ice-cold PBS (1x) for optimal conservation of tissue. Primary astrocytes were prepared from cortices of C57 BL/6 mice (postnatal day 3), or Sprague Dawley rats (postnatal day 3-5) as mixed cultures with microglia.

The handling of animals was performed according to the regulations of the German Animal Protection Law with approval from the Government of Lower Franconia, Bavaria, Germany.

After decapitation of the newborn rodents, the brain was prepared out of the skull by opening of the skullcap. The brain meninges and the cerebellum were removed, and the remaining parts were homogenized, washed and seeded in 75 cm^2 cell culture flasks in Dulbecco's Modified Eagle Medium (D-MEM) with high glutamate content. The growth medium was complemented with 10% fetal calf serum (FCS) and 1% penicillin/streptomycin. The culture flasks were coated with poly-L-ornithine (PLO). The cultures were kept at 37°C and 5% CO_2 for at least 14 days, until the astrocytes/microglia co-culture had properly differentiated, and other cell types had been abolished.

For biochemical approaches the microglia fraction of the mixed cultures was reduced by rigorous agitation of the culture flask (400 rpm) for at least 4 hours at 37°C on an orbital shaker. The medium, containing the detached microglia, was removed subsequently.

1.12 Cell treatment

For the use in experiments, the astrocytes were reseeded in glass-bottom chamberslides (for immunocytochemistry), chambered cover-glass bottom slides (for live cell imaging) or culture dishes (for biochemical assays), which were coated with poly-L-ornithine (PLO).

The cells were treated in serum-free D-MEM GlutaMax or in Leibovitz's imaging medium. Wildtype-pneumolysin (WT-PLY) was used in concentrations of 0.1-0.5 µg/ml. All experiments were carried out temperate at 37°C. Control cells (mock) were treated with an appropriate amount of pure medium.

In order to distinguish living cells from permeabilized cells, propidium iodide (PI) was added to the cells right before imaging in a 1:1000 dilution in the imaging medium or buffer. PI is able to enter cells as soon as their membrane becomes permeabilized. It emits fluorescence at around 560 nm. The total amount of cells in one imaging field was determined by adding Hoechst stain in a dilution of 1:1000 to the imaging medium or buffer at the end of an experiment. Hoechst stains the nuclei of living cells, and is excited by ultraviolet light.

Bioactive compounds were applied as follows:

Cytochalasin D: 4 µM

NSC23677 Rac1 inhibitor: 10 µM

$C3^{bot}$ exoenzyme RhoA inhibitor (ADP-ribosyltransferase): 0.1 µg/ml

Pretreatment with those components was carried out for 1 hour at 37°C.

1.13 Immunocytochemistry

Cells were re-seeded in glass-bottom cultureslides, which were coated with PLO. After treatment, the cells were washed in 1x PBS and fixed in 4% paraformaldehyde (PFA) in PBS for 30 minutes at room temperature (RT). Subsequently the cell membranes were permeabilized with 0.1% TritonX-100 for 10 minutes at RT, and unspecific binding-sites were blocked with 4% BSA in PBS for 30 minutes, also at RT. Primary antibodies were applied in concentrations, according to manufacturer recommendations, for 1 hour at RT. Goat anti-mouse or anti-rabbit F_{ab} fragments, labeled with FITC or Cy3 were used to visualize the primary antibodies. Secondary antibodies were diluted 1:200 to 1:500 in 1x PBS and applied for 1 hour at RT.

1.14 Transfection of cells

For transfection of astrocytes the Magnetofection™ approach was used following manufacturer instructions. Magnetofection™ is a suitable method to transfect cells in culture. It is based on an association of vectors with magnetic particles, which, by magnetic force, are driven towards, or even into the target cells. The magnetic nanoparticles are coated with cationic molecules. Their association with the gene vectors (DNA, siRNA, ODN, virus, etc.) is achieved by salt-induced colloidal aggregation and electrostatic interaction. With the help of the magnetic plate, the beads are targeted to the cell membrane, and thus endocytosed by the cells.

1 μg plasmid DNA was mixed with 1 μl/μl pDNA of magnetic beads and incubated in 100 μl D-MEM (w/o FCS and antibiotics) at room temperature for 20 minutes. The mixture was supplemented with a suitable amount of medium, and applied to the cells in fresh growth medium and the culture slide was positioned on a magnetic plate for 15 minutes. The cells were incubated at 37°C over night.

Plasmids used for transfection:
pTagRFP-actin
pLifeAct-TagRPF (see p. 44)

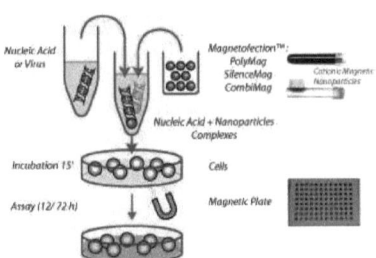

Fig. 17: *Magnetofection™ (schematic).*
The method associates nucleic acids with magnetic nanoparticles, which are coated with cationic molecules. The magnetic plate targets the beads towards the cell membrane and induces endocytosis by the cells.

1.15 Baculovirus transduction

To visualize the plasma membrane or the actin cytoskeleton of the astrocytes the Molecular Probes CelluarLights™ Baculovirus vector transfer system was used, also according to manufacturer instructions. The virus enters mammalian cells and directs the expression of auto-fluorescent proteins that are localized to specific subcellular compartments and organelles via signal peptide or protein fusion.

The virus suspension was applied in a ratio of 1:5 in Ca^{2+}- and Mg^{2+}-free D-PBS (CellularLights suspension 1 part, D-PBS 5 parts) and the cells were incubated with the mixture at 37°C for 4 hours with gentle agitation. Afterwards an enhancer solution was added to the cells and the mixture was incubated for additional 2 hours at 37°C. The solution was exchanged to growth medium and the cells were further incubated at 37°C over night.

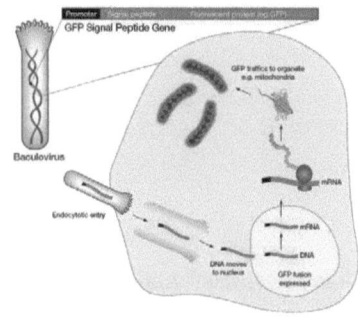

Fig. 18: ***Baculovirus transduction with Organelle Lights^{TM} (schematic).***

Organelle Lights™ reagent delivery is mediated by an insect virus (baculovirus) that is noninfectious to mammalian cells. Organelle Lights™ reagents express fluorescent protein–signal peptide fusions for accurate and specific targeting to subcellular compartments and structures, such as the plasma membrane or actin.

1.16 Microscopy

1.16.1 Bright field microscopy

Bright field microscopy is the simplest form of microscopy. Samples are illuminated with white light from below, and observed from above. This type of microscopy was used to determine sound conditions of cell cultures, also for live cell imaging, and in biochemical approaches to determine the establishment of GUVs (described below). Cell preparation was also carried out under a bright field binocular.

1.16.2 Fluorescent microscopy

Fluorescent microscopy is an essential and helpful tool in biology and biomedical science. Fluorescence is defined as the absorption and subsequent re-radiation of light by organic and inorganic specimens. The emission of light happens nearly simultaneously with the absorption of the excitation light. Therefore this form of microscopy is very suitable to investigate probes of biological origin. Through the use of multiple fluorescence labeling, different probes can identify several target molecules simultaneously. For this purpose so called fluorophores are used, which are excited by specific wavelengths of irradiating light and emit light of defined and useful intensity. Fluorescent microscopy was applied for fixed samples (immunocytochemistry), but also for time lapse imaging, imaging of fluorescently tagged actin, and for all approaches carried out on the confocal microscopes (see below).

1.16.3 Live cell imaging / time lapse imaging

Live cell experiments were carried out on a temperate Olympus Cell^M imaging fluorescent system, using a 20x dry objective or a 60x oil immersion objective. For live imaging purpose the cells were kept in Leibovitz's imaging medium, without phenol red at constant pH. Snapshots were taken at different time intervals, ranging from 5 seconds to about 2 minutes, depending on the experiment, in order to create a sequence of stacks.

1.16.4 Confocal microscopy

Confocal imaging is used for observation of either fixed, or living cells and tissues that are labeled with one or more fluorescent probes. Thick specimens, as well as cells can be easily investigated in depth by collection of serial optical sections (z-stacks). Other advantages over conventional optical microscopy are the shallow depth of field and the elimination of out-of-focus-glare. The optical detection limit of confocal microscopy is determined by the diffraction limit of light (200-300 nm). The illumination of the sample is achieved by scanning one or more focused beams of light, usually from a laser, across the specimen.

High-resolution time lapse imaging was performed on a Leica LSM SP5 using a 512x64 pixel frame, z-stacking with 10 layers and sequential scan. This kind of microscope was also used for documentation of fixed samples and for imaging of GUVs (see below).

(Based on Nikon MicroscopyU by Kenneth R. Spring and Michael W. Davidson).

1.17 Protein Biochemistry

1.17.1 Small GTPase activation pull down assays

Pull down assays were carried out following manufacturer instructions. In order to prevent previous GTPase activation by components of the culture medium (components of the FCS), the cells were starved over night in serum-free medium (D-MEM). After treatment with PLY in the starving medium for the appropriate time (5, 15, 30 minutes), the suspension was aspirated and the cells were washed once with cold 1x PBS. After aspiration of the PBS, the cells were lysed with 1 ml cell lysis buffer (50 mM Tris, pH 7.5, 10 mM $MgCl_2$, 0.5 M CaCl, 2% Igepal). The lysate was centrifuged to remove membrane fractions and the supernatant was applied to PAK-PBD for Rac1 and to Rhotekin agarose beads for RhoA, respectively. The lysate-beads suspension was rotated at 4°C for one hour, then centrifuged. The supernatant comprised the fraction, containing inactive GTPases. The beads pellet, containing the activated GTPase fraction, was washed with wash buffer (25 mM Tris, pH 7.5, 30 mM $MgCl_2$, 40 mM NaCl) for 3 times. To both fractions SDS sample buffer (containing 2.5% β-mercaptoethanol) was added (1x end concentration of the buffer), and the samples were boiled at 99°C for 3 minutes.

The pulled down activated GTPase proteins were electrophoresed on SDS-PAGE gradient gels (4-12% Bis-Tris). The separated proteins were transferred to a PVDF Imobilion-P transfer membrane via semi-dry blotting. Unspecific binding sites on the proteins were blocked using 3-5% non-fat milk in PBS-T (0.1%) and RhoA and Rac were tagged with the corresponding antibodies. After incubation with specific horseradish peroxidase-conjugated secondary antibodies, the tagged proteins were detected using ECL Plus Western Blotting reagent and a Fluorchem Q machine. Blots were evaluated via densitometric analysis of bands. In the evaluation of the GTPase activation assays, the amount of activated GTPases was equalized to total amount of GTPases in the cell lysate.

1.17.2 Cytoskeleton Isolation Assays

Primary astrocytes were seeded to 60 mm or 100 mm dishes, coated with PLO, and grown to a dense cell monolayer in culture medium (D-MEM, 10% FCS, 1% Pen/Strep). After changing the medium to pure D-MEM, the cells were challenged with 0.2 µg/ml of toxin for 10 minutes. The cells were washed with cold PBS and subsequently harvested with the help of a cell scraper in cell lysis/cytoskeleton stabilization buffer (50 mM PIPES, pH 6.9, 5 mM $MgCl_2$, 5 mM EGTA, 20% v/v Glycerol, 0.2% v/v Nonidet P40, 0.2% v/v Triton X-100, 0.2% v/v Tween 20, 0.01% 2-mercapto-ethanol, 0.001% Antifoam C; pre-warmed to 37°C). In order to achieve complete lysis, the cells were further homogenized with a 27G syringe. When performing microtubule reducing experiments, $1/100^{th}$ volume of nocodazole was added. For F-actin reduction $1/200^{th}$ volume of gelsolin and $1/100^{th}$ volume of cytochalasin D were added. For degradation of both cytoskeleton components, all three agents were added. The samples were incubated at 37°C for 30 minutes, and subsequently ultracentrifuged at 50.000 x g for 1 hour at 37°C. The supernatants were kept on ice after centrifugation. The pellet fractions were resuspended in ice cold MilliQ water for 1 hour, for complete dissociation of cytoskeleton components. 10 μM CytoD were added for acceleration of the process. The samples were diluted in SDS-sample buffer (containing 2.5% β-mercaptoethanol), boiled at 99°C for 3 minutes, and electrophoresed on SDS-PAGE gradient gels (4-12% Bis-Tris). The separated proteins were transferred to PVDF Imobilion-P transfer membranes via semi-dry blotting. Unspecific binding sites on the proteins were blocked using 3-5% non-fat milk in PBS-T (0.1%) and PLY was tagged with a specific PLY antibody. After incubation with a specific horseradish peroxidase-conjugated secondary antibody, the probed PLY was detected using ECL Plus Western Blotting reagent and a Fluorchem Q machine.

1.17.3 Preparation of G-actin and F-actin

In order to obtain G-actin, non-muscle actin, isolated from human platelets (>99% pure), was reconstituted in a buffer containing 5 mM Tris-HCl, pH 8.0, 0.2 mM $CaCl_2$, 0.2 mM ATP, 5% (w/v) sucrose, and 1% (w/v) dextran. In order to obtain F-actin, 1/10 actin polymerization buffer (500 mM KCl, 20 mM $MgCl_2$, 10 mM ATP) was added to the G-actin, and the solution was incubated at 24°C for 1 hour. Preformed F-actin was reconstituted in a general actin buffer (5 mM Tris-HCl pH 8.0, 0.2 mM $CaCl_2$).

1.17.4 Actin binding assays

Freshly prepared F-actin (see above) (4.6 µM) together with PLY and its variants (2.5 µM) was incubated for 1 hour at 24°C at a total volume of 100 µl, subsequently centrifuged at 150,000 x g. The pellet, containing the actin was resuspended in 50 µl MilliQ water. SDS sample buffer (containing 2.5 b β-mercaptoethanol) was added to both fractions and the samples were boiled at 99°C for 3 minutes, and subsequently electrophoresed on SDS-PAGE gradient gels (4-12% Bis-Tris). The gels were either directly stained with Coomassie blue protein stain, or the separated proteins were transferred to PVDF Imobilion-P transfer membranes via semi-dry blotting. Unspecific binding sites on the proteins were blocked using 3-5% non-fat milk in PBS-T (0.1%) and PLY was tagged with a specific PLY antibody. After incubation with a specific horseradish peroxidase-conjugated secondary antibody, the probed PLY was detected using ECL Plus Western Blotting reagent and a Fluorchem Q machine.

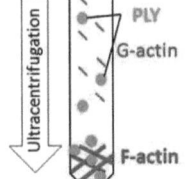

Fig. 19: *Actin Spin Down Assay (schematic).*

1.17.5 Enzyme-linked sorbent assays (ELSAs)

F96 Cert. Maxisorp Immuno plates were coated with PLY variants (20 µg/ml) overnight at 4°C. Unspecific binding sites were blocked with 10% fetal calf serum (FCS) for 1 h. G-actin was added in a concentration of 20 µg/ml, Arp2/3 protein complex in a concentration of 4 µg/well (80 µg/ml). After 2 hours incubation at room temperature (RT), the wells were washed 3 times with 1x PBS-T 0.05% (PBS, Tween-20). Anti-actin or anti-Arp antibodies (1:1000 or 1:500 in PBS) were incubated for 1 h at room temperature. HRP-conjugated secondary antibody was applied for 1 h in a 1:1000 dilution in PBS, and the reaction was visualized using 50 µl/well TMB substrate solution (30 minutes incubation). The reaction was stopped with 50 µl/well of 0.18 M H_2SO_4. Substrate absorption was measured at 490 nm using an EL800 ELISA plate reader.

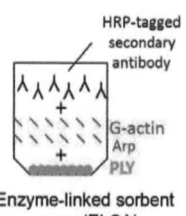

Enzyme-linked sorbent assay (ELSA)

Fig. 20: ***ELSA assay (schematic).***
Maxisorp Immuno plates were coated with PLY or its variants/fragments. Subsequently the amount of bound actin or Arp was detected with the help of an HRP-tagged antibody and measurement of absorption.

1.17.6 Actin-pyrene polymerization assays

One vial of 10 mg non-muscle platelet actin and the same amount of pyrene actin were resuspended in 50 µl double-distilled (MilliQ) water. The viscous solution was mixed (10% pyrene actin) and diluted in general actin buffer (5 mM Tris-HCl, pH 8.0, 0.2 mM $CaCl_2$) to a final actin concentration of 9 µM. The actin suspension was centrifuged at 150.000 x g for 1.5 hours at 4°C to remove pre-formed actin aggregations. The supernatant was further diluted in general actin buffer to a final actin stock concentration of 3 µM.

Polymerization of actin was measured by detection of actin-pyrene fluorescence in a fluorescence photometer (Luminescence Spectrometer), using a cuvette and an actin stock volume of 600 µl.

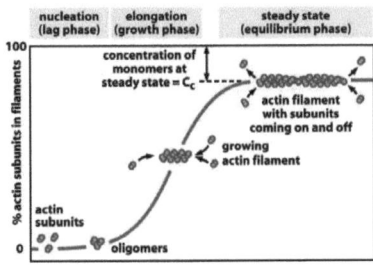

Fig. 21: Time course of in vitro actin polymerization in the absence of effector proteins.
(Source:[Alberts et al., 2008])*

Parameters for measurement:

Excitation: 350 nm

Emission: 407 nm

Interval: 10 sec

Cycles: 300 - 1800

Integration time: 0.10 sec

Excitation slit: variable (~ 5.0)

Emission slit: variable (~ 6.5)

(Single read and timedrive modus)

Software: FL WinLab (Perkin Elmer)

After measuring background fluorescence of actin for at least 5 minutes, additional components were added directly to the actin suspension.

Concentrations of components:

Actin polymerization buffer: 1x (10x stock)

Arp2/3 protein complex: 50 nm

Pneumolysin: variable

VCA: 50 nm

Inhibitory PLY-peptides: 10-fold concentration of PLY

1.17.7 Imaging of actin-rhodamine filaments

Non-muscle actin, actin-rhodamine (actin human TRITC) and actin-biotin were diluted to a final stock concentration of 1 mg/ml in double-distilled water (MilliQ) and mixed in a ratio of 65% : 25% : 10% (23 µM). Actin polymerization buffer was added to the actin mixture in a 1x final concentration. The mixture was incubated at 30°C for 1 hour for pre-formation of actin filaments. The 23 µM actin stock was subsequently diluted to a final stock concentration of 200 nm in TIRF-buffer.

TIRF-buffer:

50 mM KCl

1 mM $MgCl_2$

1 mM EGTA

10 mM imidazole

2 mM ATP

0.2 mM β-mercaptoethanol

4.5 mg/ml glucose

4.3 mg/ml glucose oxidase

0.7 mg/ml catalase

Coverslip chamberslides or ibidi µ-slides were coated with 0,25 mg/ml NeutrAvidin® (streptavidin tag) for 2 minutes, in order to immobilize the actin-biotin containing actin fibers. NeutrAvidin® (5 mg/ml stock) was diluted in F-buffer.

F-buffer:

50 mM KCl

1 mM $MgCl_2$

1 mM EGTA

10 mM imidazole

2 mM ATP

0.2 mM β-mercaptoethanol

Unspecific binding sites were subsequently blocked for 2 minutes with 1 mg/ml BSA (bovine serum albumin), which was diluted in F-buffer from a 10 mg/ml stock. After blocking, 30 µl of the 200 nm actin mix were added to the coated slide. After 2 minutes the unbound actin was removed by rinsing the site with TIRF-buffer, and the attached filaments were imaged immediately using fluorescent microscopy.

Pneumolysin was used in a 1 µM end-concentration. For calcium supplementation PLY was pre-incubated with 2 mM Ca^{2+} for 30 minutes at room temperature.

VCA domain protein and Arp2/3 protein complex were used in a 60 nM end-concentration.

1.17.8 Far Western Blotting (Overlay Blot)

Actin (20 µg non-muscle platelet actin protein) and Rho (20 µg recombinant protein) were mixed with an equivalent amount of SDS-sample buffer, containing 2.5% β-mercaptoethanol, and boiled at 99°C for 3 minutes. The denatured proteins were electrophoresed on a 4-12% Bis-Tris gradient gel at 120 V for one hour. Subsequently the proteins were blotted onto a PVDF membrane by using an iBlot® Dry Blot device (program 5, 7 minutes). After blotting the membranes were incubated in binding buffer (50 mM NaH_2PO_4, pH 7.5, 0.3% Tween-20) for 30 minutes. The control membrane was further incubated in binding buffer for 2 hours, whereas the overlay-membrane was incubated in binding buffer containing 0.5 µg/ml PLY. Subsequently both membranes were washed 6 times in 0.1% PBS-T for 5 minutes each time. Unspecific binding sites on the proteins were blocked in 3% skimmed milk (in 0.1% PBS-T), and then incubated with a specific PLY antibody (1:1000) in PBS. Incubation with a HRP-conjugated goat anti-mouse IgG (1:1000 in PBS) followed after a 3 times washing step, 10 minutes each. Detection of the overlay was carried out on a FluorChemQ gel detection device, using ECL Plus Western Blot Detection reagent.

1.17.9   FRET experiments

Förster/Fluorescence resonance energy transfer (FRET) is able to reveal the proximity of two fluorescently labeled molecules over distances > 100 Å (1 - 10 nm), and thus can be used to investigate the affinity, binding or interaction between two proteins [Kenworthy, 2001].

Here FRET was used to show the co-localization of Atto488-tagged PLY and phalloidin-Alexa555, using a luminescence spectrometer (photometer). The PLY-Atto488 stock was diluted 1:50 in PBS, resulting in a 5 µg/ml concentration of the fluorescence donor. Phalloidin-Alexa555 was added in gradually dilutions (1:2.5, 1:3.3, 1:3.8) and the fluorescence energy transfer was measured between 500 and 600 nm.

1.18 Giant unilamellar vesicle (GUV) approach

GUVs comprise vesicles of 1-30 µm in diameter, consisting of a lipid bilayer, composed of a defined set of phospholipids [Baumgart et al., 2003; Colletier et al., 2002; Varnier et al., 2010]. In this case the GUVs were composed of DOPC (1,2-Dioleoyl-sn-glycero-3-phosphocholine), DPPC (1,2-dipalmitoyl-sn-glycero-3-phosphocho- line), and cholesterol in a ratio of 25% DOPC : 35% DPPC : 40% cholesterol. All components were diluted in chloroform. In order to obtain solvent-free membranes, 20 µl of the lipid solution were applied to a coverslip, and the dissolvent was removed completely by vacuum evaporation in an exsiccator. A ring, treated with grease, was placed around the dried lipids and then filled with D-sorbitol solution. In the Vesicle Prep Pro® device (Nanion), the GUVs were formed through electro-swelling (hydration of a dry lipid film in an oscillating electric field).

Liposomes, containing GFP-tagged PLY were composed of 40% DOPC and 60% DPPC. The lipids were dried at the bottom of an 8 ml tube for about 2 hours. Multi-lamellar vesicles were obtained by hydration of the film with 0.1 ml of 25 mM MOPS buffer, pH 8.5, containing 0.3 nmol of the toxin. The tube was vortexed until the lipid film had peeled off from the tube surface. To break the multi-lamellar vesicles into mono-lamellar, ten cycles of freezing (liquid nitrogen) and thawing (30°C water bath) were applied. The sample was then diluted to 1 ml in 25 mM MOPS buffer. The size of liposomes was homogenized by extrusion by passing the sample 10-fold through a 200 nm pore polycarbonate filter, or by centrifugation. Carriage of PLY-GFP from the liposomes into the GUVs was achieved by cold-fusion (4°C over night).

Depletion of extra-vesicular PLY-GFP was achieved by incubation of the liposome/PLY-GFP solution with cholesterol containing liposomes (composed of DOPC, DPPC). Free, non-encapsulated PLY was captured by incubation for 2 hours on a shaker platform with a cholesterol-containing trap, which consisted of a dried multi-lamellar film of cholesterol-containing lipid mixture as was used for the GUVs, and centrifuged to eliminate all big lipid sheets containing the bound PLY and stained fluorescent green (due to PLY-GFP). The liposome mixture was tested for cytotoxicity and toxin binding, which demonstrated a complete lack of free PLY. The fused GUVs demonstrated proper PLY-GFP staining localized on the surface of the GUV. To verify the orientation of the PLY-GFP on the GUVs' surface (which was expected to be on the inside, and thus the GFP should be hanging inside due to its tagging to the C-terminal of the toxin), we incubated GUVs with a strong quencher (trypan blue). The trypan blue should have quenched all the GFP fluorescence given that the toxin binds from the outside of the GUV, but the GFP fluorescence remained unchanged.

For actin polymerization experiments, the GUVs with or without PLY were incubated with rhodamin-labeled G-actin in a final mix containing 7.5 µM G-actin, 160 nM Arp2/3 complex and 1 mM ATP/Mg^{2+} containing actin polymerization buffer [Colletier et al., 2002].

Actin-TRITC was prepared freshly, by dilution of 10 µg in 30 µl ddH$_2$O (MilliQ). The imaging sample, which was applied to an ibidi µ-slide, was composed of: 22 µl aqua purificata

15 µl actin-TRITC

5 µl 10x actin polymerization buffer

8 µl Arp2/3 protein complex (0.5 µg/µl stock)

Imaging was performed on the Leica SP5 confocal microscope.

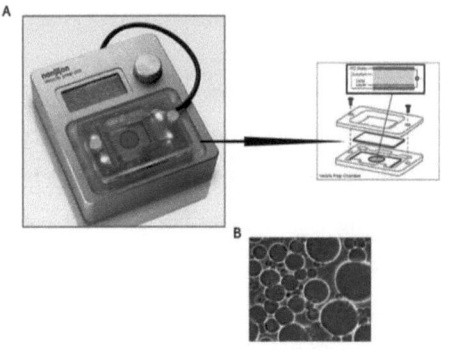

Fig. 22: ***Nanion Vesicle Prep Pro®.*** *Generation of an electric field in the Vesicle Prep Pro device (A) leads to electro-swelling of the dried phospholipid mixture in a D-sorbitol solution, resulting in the formation of so called giant unilamellar vesicles (GUVs), with a diameter of 10-30 µm (B). (Source: www.nanion.de; 2011)*

1.19 Cell border tracking assay

In order to properly evaluate pneumolysin effects on the cell shape and movement behavior of astrocytes, a cell border tracking assay, using ImageJ software (version 1.43 for Windows, National Institute of Health, Bethesda, Maryland, U.S.), was established. After sequential imaging of astrocytes (live cell imaging approach), the pictures were processed with ImageJ and then one single spot of the cell border of an astrocyte was tracked throughout the whole image stack from time point zero to the end with the „manual tracking" tool.

Subsequently the length of the line was measured with the help of the "line" tool, comprising the distance of cell movement. Per field, at least 10 cells were evaluated with this method. The mean distance in µm and the SEM were calculated. Treated cells were compared to mock treated cells.

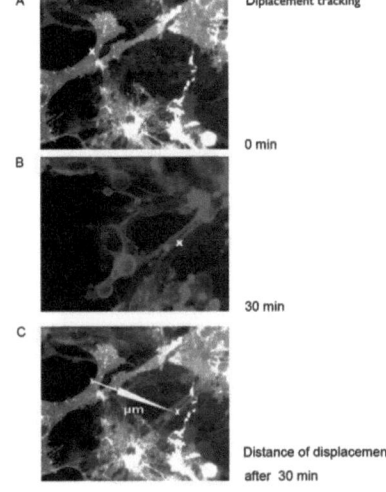

Fig. 23: *Displacement tracking of living cells.*

*The cell border of a single cell (evaluation of ~10 cells/field) was tracked throughout a whole time stack from 0 min (white mark in **A**) to the end (white mark in **B**) with the help of the manual tracking plugin of the ImageJ software, and the distance in µm was determined (C).*

*(**C**) shows the merge of the first (grey) and last (red) image of the time stack and the spatial shift of the cell (arrow).*

1.20 Preparation and resealing of erythrocyte ghosts

Erythrocyte ghost were prepared as described in [Prausnitz et al., 1993] with some modifications. Shortly, we utilized fresh human donor blood, incubated with EDTA to avoid coagulation. Following lysis with 5 mM PBS buffer, pH 8.5, the ghosts were thoroughly washed and loaded with 2 mM calcein in 5 mM PBS buffer, pH 8.5 for 30 minutes on ice, followed by 1 hour additional incubation with 25 µg/ml bovine Arp2/3 or 25 µg/ml rabbit non-specific purified IgG as a negative control. The ghosts were resealed in 40 mM PBS pH 7.4 at room temperature for 2 hours, thoroughly washed and incubated in the same buffer with PLY. Following incubation with PLY for 20 minutes at 37°C, the remaining intact resealed ghosts were pelleted and lysed and their calcein fluorescence compared with the corresponding control, not exposed to PLY. The concentration of PLY was titrated to achieve ~40-50% lysis within 20 minutes of 1% (v/v) red blood cell ghosts. The identical shape and size of all resealed ghosts was confirmed by confocal microscopy, as well as the complete loss of their calcein fluorescence upon lysis by PLY.

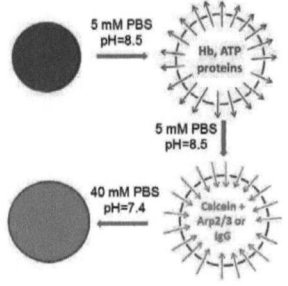

Fig. 24: ***Schematic of erythrocyte preparation.***
Schematic presentation of the erythrocyte ghost preparation, including erythrocyte lysis, loading with calcein and Arp2/3 or IgG and subsequent resealing.

1.21 Graphical and statistical analysis

Graphical analysis was carried out with ImageJ, version 1.44f for Windows (National Institute of Health, Bethesda, MD, U.S.).

Evaluations and statistical analysis were carried out with MS Excel 2007 and with GraphPad Prism, version 4.02 for Windows (Graph Pad Software, Inc.).

*: "Copyright 2008 from Molecular Biology of the Cell by Alberts et al. Reproduced by permission of Garland Science/Taylor & Francis Books, LLC."

1.22 Equipment and Materials

1.22.1 Pneumolysin and plasmids

Pneumolysin and all its fragments and mutants, as well as their GFP-tagged variants (all diluted in PBS) were friendly provided by Professor Timothy J. Mitchell and Jiangtao Ma at the University of Glasgow, Scotland. Atto488 labeling of pneumolysin was performed by Christina Förtsch, University of Würzburg, Germany.

Wild-type toxin:
- **PLY-WT**
- PLY-eGFP
- PLY-Atto488

Mutants:
- **Δ6-PLY** (Δ A146 R147, mutation in the unfolding domain 3)
- Δ6-eGFP
- **W433F-PLY** (Trp433→Phe mutation in the Trp-rich loop of domain 4)

Fragments:
- **D123-PLY** (domains 1, 2 and the unfolding domain 3)
- **D4-PLY** (binding-domain 4)

Plasmids	Source
pLifeAct-TagRFP	ibidi

p^{CMV} LifeAct-TagRFP vector, 4.7kb

P$_{CMV IE}$ — LifeAct — TagRFP — SV40 poly A — P$_{SV40}$ — Neo^R — HSV TK poly A

| pTagRFP-actin | Evrogen |

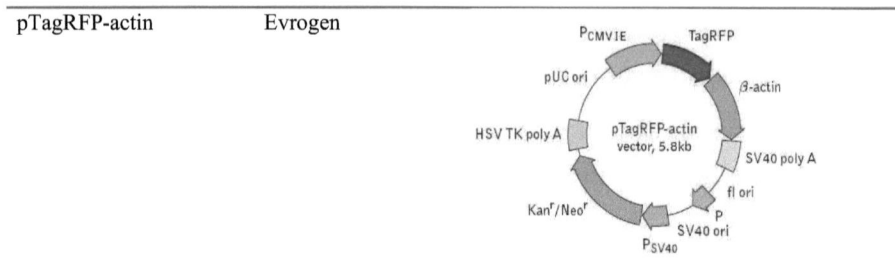

1.22.2 Materials, animals, equipment and software

Devices/Material	Manufacturer
Autoclaves	Systec (V 150), HMC (HV-110)
Balance	Kern
Centrifuges (table top)	Hettich, Eppendorf
Chambered coverslips and coverglasses, µ-slides	Nunc, BD Falcon, ibidi
Exsiccator	Roth
ELISA plate reader	BioTek (EL 800)
Freezer	Heraeus (Hera freeze, -80°C)
Gel/WB detector	FluochemQ, AlphaInnotech
Glass cuvette	Hellma Analytics
Ice machine	Scotsman (AF 200)
Incubators	Heraeus (Hera Cell 240), Memmert
Instruments for cell preparation	Roth, Hartenstein
Labware (Cell culture flasks, pipettes, multi-well plates, bottle-top filters, etc.)	Sarstedt, Roth, StarLab, Eppendorf, Millipore, BD
Laminar flow cabinets (sterile)	BDK Luft- u. Reinraumtechnik
Maxi Sorp Plates	Nunc, BD Falcon
Orbital shaker	Biometra (WT 12), NeoLab
pH meter	Hanna Instruments
Photometer	Perkin Elmer Instruments, Luminescence Spectrometer LS 50B
Pipettes	Eppendorf, Gilson
Plastic cuvettes	Roth
Power supply units	Biorad (Power Pac 3000), Consort (EV 231)
Precision Balance	Sartorius
Protein electrophoresis chambers	Invitrogen (Life Technologies)
PVDF membrane	Millipore (Immobilon-P)
Refrigerators	Liebherr
Rotator	Stuart (SB2)
Stirrer	Heidolph

Syringes	Braun
Thermomixer	Eppendorf
Ultra centrifuge tubes	Beckmann, Herolab
Ultra centrifuges	Beckmann
Vacuum pump	IlmVac
Vesicle Prep Pro (GUV device)	Nanion Technologies
Water baths	Memmert

Chemicals/Media/Proteins	**Manufacturer**
Alcohols, buffers, milk powder, acids, bases, DMSO, other chemicals	Roth, Sigma, AppliChem, Sigma-Aldrich
Aqua purificata	Roth
Cell Culture Media, D-PBS	GIBCO, Invitrogen (Life Technologies)
Cholesterol	Sigma-Aldrich
Coomassie® Brilliant Blue R250 Powder	Serva Electrophoresis
DOPC, DPPC (phospholipids)	Echelon Inc.
Fetal calf serum	Pan Biotech
Chromatography columns	Bio Rad
HEPES buffer	GIBCO, Invitrogen
LPS (Endotoxin)	Lonza
MilliQ water (double distilled water)	Millipore filter unit
Mowiol 4-88	Roth
NeutrAvidin® (avidin)	Invitrogen, Molecular Probes (Life Technologies)
Paraformaldehyde	Roth
Penicillin/Streptomycin (liquid solution)	GIBCO, Invitrogen (Life Technologies)
Poly-L-lysine (PLO)	Sigma-Aldrich
Protease Inhibitor Cocktail	Roche
TMB substrate solution	Pierce (ThermoScientific)

Microscopes	**Manufacturer**
Binocular	Motic Microscopes
Bright-field light microscopes	Leica, Motic Microscopes
Fluorescent microscope (wide field, inverted)	Olympus
SP5 confocal microscope	Leica

Laboratory animals	**Provider**
C57BL/6 mice	Janvier, France
Sprague Dawley rats	Janvier, France; CharlesRiver, Germany

Primary Antibodies	Manufacturer
Anti-actin (mouse monoclonal), IgG$_1$	Santa Cruz
Anti-Arp2 (rabbit polyclonal), IgG	Abcam
Anti-Arp3 (mouse monoclonal), IgG$_1$	Santa Cruz
Anti-GFAP (rabbit polyclonal)	Zymed® Laboratories (Invitrogen Immunodetection)
Anti-p21-ARC (rabbit polyclonal), IgG	Santa Cruz
Anti-p41-ARCb (rabbit polyclonal), IgG	Santa Cruz
Anti-PLY (mouse monoclonal), IgG$_1$, IgG2b	Santa Cruz, Abcam
Anti-Rac (rabbit polyclonal)	Cytoskeleton Inc.
Anti-Rac1/2/3 (rabbit polyclonal), IgG	Cell Signaling
Anti-Rho (mouse monoclonal)	Cytoskeleton Inc.
Anti-Vinculin hVIN-1 (mouse monoclonal), IgG$_1$	GeneTex Inc.

Secondary Antibodies	Manufacturer
Cy^{TM}3-conjugated AffiniPure F(ab')$_2$ fragment goat anti-mouse IgG	Jackson Immuno Research Laboratories Inc.
Cy^{TM}3-conjugated AffiniPure F(ab')$_2$ fragment goat anti-rabbit IgG	Jackson Immuno Research Laboratories Inc.
Fluorescein (FITC)-conjugated AffiniPure F(ab')$_2$ fragment goat anti-rabbit IgG	Jackson Immuno Research Laboratories Inc.
Fluorescein (FITC)-conjugated AffiniPure F(ab')$_2$ fragment goat anti-mouse IgG	Jackson Immuno Research Laboratories Inc.
Peroxidase-conjugated AffiniPure goat anti-mouse IgG	Jackson Immuno Research Laboratories Inc.
Peroxidase-conjugated AffiniPure goat anti-rabbit IgG	Jackson Immuno Research Laboratories Inc.

Actins	Manufacturer
Actin-biotin	Cytoskeleton Inc.
Actin-pyrene (pyrene muscle actin)	Cytoskeleton Inc.
Actin-TRITC (rhodamine)	Cytoskeleton Inc.
Non-muscle actin (human platelet)	Cytoskeleton Inc.
Pre-formed actin filaments (rabbit skeletal muscle)	Cytoskeleton Inc.

Commercial Kits	Manufacturer
Actin Binding Protein Biochem Kit	Cytoskeleton Inc.
Actin Polymerization Biochem Kit	Cytoskeleton Inc.
Active Rac1 Pull-Down and Detection Kit	ThermoScientific

BCA Protein Assay Kit	Pierce (ThermoScientific)
Cytoskeleton Isolation and Assay Kit	Cytoskeleton Inc.
Limulus Amebocyte Lysate Kit	Lonza
Magnetofection™ PolyMag	Oz Biosciences
Rac Activation Assay Kit	Cell Biolabs Inc.
Rac1 Activation Assay Biochem Kit	Cytoskeleton Inc
Rho Activation Assay Biochem Kit	Cytoskeleton Inc.

SDS-PAGE equipment	Manufacturer
Gels	
NuPAGE 4-12% Bis-Tris Gels	Invitrogen (Life Technologies)
Anamed VarioGel 4-12%	Anamed
NuPAGE MOPS SDS buffer	Invitrogen, Anamed
NuPAGE Sample loading buffer	Invitrogen, Anamed
NuPAGE Antioxidant	Invitrogen
NuPAGE Sample reducing agent	Invitrogen
NuPAGE Transfer buffer	Invitrogen
Molecular weight markers:	
SeeBlue® Plus Prestained Standard 1x	Invitrogen
PAGE Ruler Prestained Protein Ladder	Fermentas
Biotinylated Protein Ladder	Cell Signaling
Novex® Semi-Dry Blotting Device	Invitrogen
iBlot® Dry Blot Device	Invitrogen
iBlot® Gel Transfer Stacks PVDF Mini	Invitrogen
ECL Plus Western Blotting Detection System	GE Healthcare, Amersham™

Components for actin assays	Manufacturer
Actin Polymerization Buffer (10x)	Cytoskeleton Inc.
Arp2/3 protein complex (bovine brain)	Cytoskeleton Inc.
ATP (Adenosine tri-phosphate)	Cytoskeleton Inc.
General Actin Buffer	Cytoskeleton Inc.
WASP-VCA Domain-GST protein	Cytoskeleton Inc.

Cell active components/Inhibitors/fluorophores	Manufacturer
Atto488, Atto561	Atto Tec GmbH
$C3^{bot}$ Transferase Protein (bacterial recombinant) Rho inhibitor	Cytoskeleton Inc.

Calcein	Invitrogen, Molecular Probes
Cellular Lights™ Intracellular Targeted Fluorescent Proteins	Invitrogen, Molecular Probes (Life Technologies)
Baculovirus PM-GFP	
Baculovirus Actin-GFP	
Cytochalasin D	Calbiochem
Hoechst Stain	Invitrogen, Molecular Probes
NSC23677 Rac1 inhibitor	Calbiochem
Phalloidin-Alexa Fluor® 488	Invitrogen, Molecular Probes
Phalloidin-Alexa Fluor® 555	
Propidium Iodide Nucleic Acid Stain	Invitrogen, Molecular Probes

Inhibitory peptides	Manufacturer
Peptide I YGQVNNVPAR	PanaTecs
Peptide II MQYEKITAHS	PanaTecs
Peptide III MEQLKVKFGS	PanaTecs
Peptide IV RPLVYISSVA	PanaTecs
Peptide V YGRQVYLKLE	PanaTecs
Peptide VI MANKAVNDFILAMNYDKKKLLTHQG-ESIENRFIKEGNQLP	Caslo Laboratories

Application	Software
ELSA assays	Gen5™ (BioTek)
Fluorescence photometer measurements	WinLab (Perkin Elmer)
Image analysis	ImageJ (NIH)
Imaging	Cell^M (Olympus), LAS AF (Leica)
Statistics, Evaluations	GraphPad, Inc. (Prism), MS Office
Western Blot detection	AlphaView® (AlphaInnotech)

1.22.3 Buffers and solutions

Phosphate buffered saline 10x (PBS):
1368 mM NaCl
26.8 mM KCl
101.4 mM Na_2HPO_4
17.6 KH_2PO_4
pH 7.3

PBS-T 0.1%:
1x PBS
0.1% Tween-20

Blocking solution (Western Blots):
PBS-T 0.1%
3-5% skimmed milk powder

Coomassie blue staining solution:
0.1% brilliant blue
50% methanol
10% acetic acid
40% dH_2O

Paraformaldehyde fixing solution:
450 ml dH_2O at 60°C
20 g PFA
5 drops 2N NaOH
50 ml 10x PBS
pH 7.2

Poly-L-ornithine coating solution:

0.15 M (3.71 g) boric acid

0.025 mg/ml (100 mg) PLO

Mowiol 4-88 mounting solution:

2.4 g Mowiol

6 ml ddH_2O

6 g glycerol

12 ml 0.2 M Tris-HCl pH 8.5

2.5% DABCO

Calcium-free imaging buffer:

135 mM NaCl

2 mM $MgCl_2$

4 mM KCl

5 mM HEPES

pH 7.3

Calcium-containing imaging buffer:

135 mM NaCl

2 mM $MgCl_2$

2 mM $CaCl_2$

4 mM KCl

5 mM HEPES

pH 7.3

Binding buffer (Overlay blotting):

50 mM NaH_2PO_4, pH 7.5

0.3% Tween-20

Cell lysis buffer:
50 mM Tris, pH 7.5
10 mM $MgCl_2$
0.5 M CaCl
2% Igepal

Wash buffer:
25 mM Tris, pH 7.5
30 mM $MgCl_2$
40 mM NaCl

General actin buffer:
5 mM Tris-HCl pH 8.0
0.2 mM $CaCl_2$

Actin polymerization buffer:
500mM KCl
20mM $MgCl_2$
10mM ATP

TIRF buffer:
50 mM KCl
1 mM MgCl2
1 mM EGTA
10 mM imidazole
2 mM ATP
0.2 mM β-mercaptoethanol
4.5 mg/ml glucose
4.3 mg/ml glucose Oxidase

0.7 mg/ml catalase

F-buffer:

50 mM KCl

1 mM MgCl2

1 mM EGTA

10 mM imidazole

2 mM ATP

0.2 mM β-mercaptoethanol

Cell lysis/cytoskeleton stabilization buffer:

50 mM PIPES pH 6.9

5 mM $MgCl_2$

5 mM EGTA

20% v/v glycerol

0.2% v/v Nonidet P40

0.2% v/v Triton X-100

0.2% v/v Tween 20

0.01% 2-mercaptoethanol

0.001% Antifoam C

2 RESULTS

2.1 Determination of the astroglial nature of cells

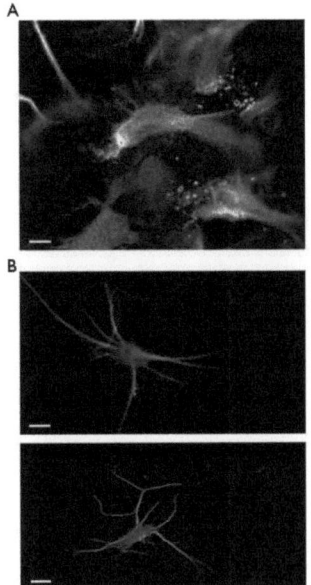

Fig. 25: **Primary mouse astrocytes.**
A. GFAP-staining shows the astrocytic nature and abundance of astrocytes in the culture.
B. GFAP-staining also reveals the star-shaped phenotype of astrocytes, which partly bear long processes. Scale bars: 10µm; anti-GFAP staining of fixed astrocytes, visualized by epi-fluorescent microscopy; 20xmagn.

After isolation of astrocytes from mice or rat brains, the cells were cultured for two weeks in flasks and for investigation they were re-seeded in chambered cover-glasses or chamberslides (all coated with Poly-L-ornithine). In order to determine the astroglial nature of the cells in use, the Glial Fibrillary Acidic Protein (GFAP), which is a specific component of intermediate filaments of astrocytes, was visualized by tagging with a specific anti-GFAP antibody in PFA-fixed cells (immunocytochemistry).
Fluorescent microscopy of the fixed samples revealed a strong GFAP staining and abundance of GFAP in the majority of cells (Fig. 25 A), which also showed typical

characteristics of astrocytes, e.g. star-shape. Comparison of Fig. 25 with Fig. 15 reveals that astrocytes adapt different phenotypes in brain tissue (Fig. 15) and in culture (Fig. 25). Whereas astrocytes appear highly branched when connected with neurons, in culture they develop a broad phenotype (Fig. 25 B) leading to establishment of an astrocytic monolayer (as depicted in Fig. 26/Fig. 28).

2.2 Definition of sub-lytic concentrations of pneumolysin (PLY)

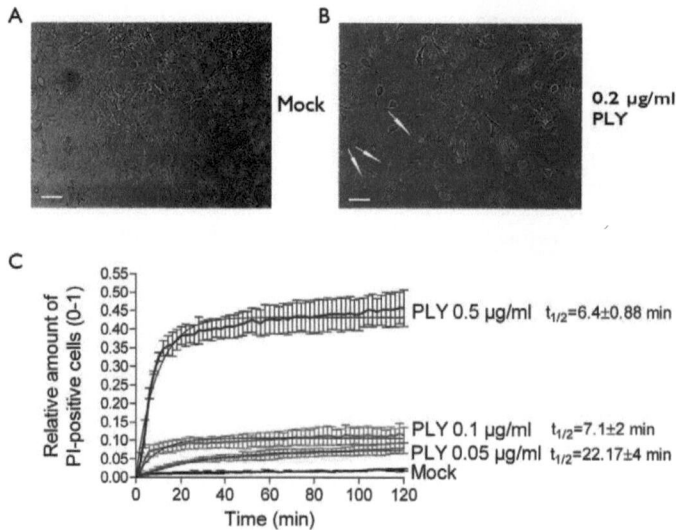

Fig. 26:   *Lytic capacity of PLY.*
A/B. Permeabilization of cells judged by propidium iodide (PI) staining. A high number of PI-positive cells represent microglia as indicated by the white arrows. The permeabilized cells are round and immobile. Scale bars: 20 μm.
C. Permeabilization curves of astrocytes following challenge with different concentrations of PLY (0.05 - 0.5 μg/ml); n=3 experiments, the values represent mean ± SEM.

This study was set to investigate the effects of the bacterial protein toxin pneumolysin on primary astrocytes under physiological conditions. In the liquor of patients suffering from pneumococcal meningitis, concentrations of PLY range between 0.001 and ~0.2 µg/ml [Spreer et al., 2003]. As this study used *in vitro* and not *in vivo* conditions, it had to be determined which concentrations of the toxin had lytic effects on the astrocytes. In this context it was only distinguished between lytic (permeabilized) and non-lytic (non-permeabilized), as a transient macro pore state could not be excluded. Therefore propidium iodide was added to mock-treated and PLY-treated cells, and the number of permeabilized cells within a time lapse was determined.

The astrocytes were challenged with increasing concentrations of PLY (0.05-0.5 µg/ml). Independent of the concentration, the permeabilization of cells occurred in a time lapse of 0 to 40 minutes. With an increase of the concentration the halftime of permeabilization dropped from ~22 minutes (0.05 µg/ml) to ~6 minutes (0.5 µg/ml). At a concentration of 0.1 µg/ml of PLY, 90% of cells were intact after 40 minutes from the time point of toxin application. Additionally many of the PI-positive cells were identified as microglia, which crawl on top of the astrocytic layer and have a clearly different morphology and a higher motility compared to the astrocytes (Fig. 26). Dead cells demonstrated a lack of motility and a rounded morphology. At higher concentrations (0.5 µg/ml) also many of the astrocytes in the monolayer were PI-positive, with rounded cell bodies and without processes [Förtsch et al., 2011].

For all further investigation of PLY effects on primary astrocytes a toxin concentration of 0.1 µg/ml (first toxin charge) and later 0.2 µg/ml (second toxin charge) was used (see Method part, chapter 2.2), latter in order to guarantee an equivalent rate of active toxin molecules within the treatment solution.

2.3 Investigation of effects of sub-lytic PLY concentrations on astrocytes and the cytoskeleton

Astrocytes were seeded on chambered coverglasses in a density providing the cells the possibility to build up inter-cellular contacts and a loose monolayer. The cells were investigated using either transmission light imaging or fluorescent imaging, latter after transfection or transduction with fluorescent proteins or probes.

Observation of control cells revealed minimal movement, with minimal displacement in a certain time frame. Slight membranous changes and fluctuations could be observed on the quiescent cells (Fig. 27).

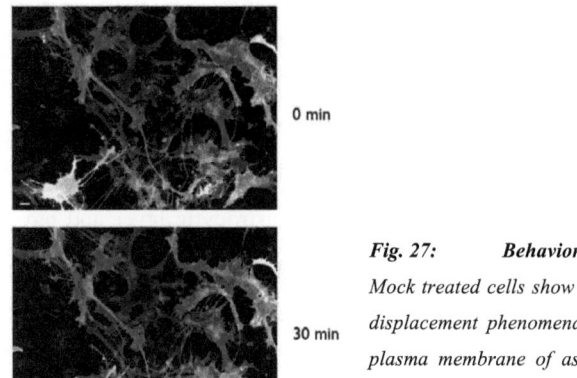

Fig. 27: **Behavior of mock-treated astrocytes.**
Mock treated cells show no or slightly cell shape changes or displacement phenomena as judged by GFP-staining of the plasma membrane of astrocytes and subsequent time lapse imaging. Scale bars: 10 μm fluorescence microscopy, 20xmagn.

Application of 0.2 μg/ml PLY to the cells led to distinct morphological changes and to changes in the cell position. The contraction of the cell body induced cell displacement (up to 7 μm in distance from the original position), distraction, and disruption of the monolayer (Fig. 28 A).

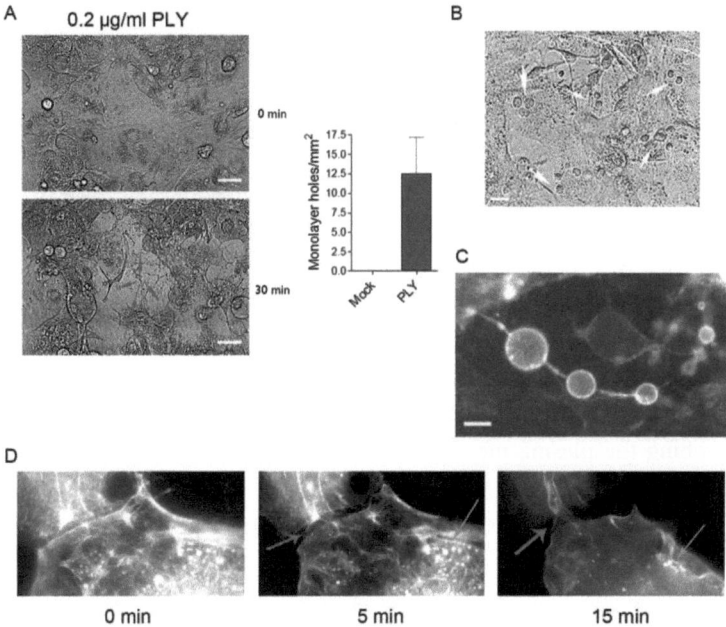

Fig. 28: **Effects of sub-lytic amounts of PLY (0.2 µg/ml) on primary astrocytes.**

A. Challenge of the cells with PLY leads to severe cell shape remodeling. The shrinkage and displacement of cells induces the formation of big holes in the cell monolayer (indicated by the red arrows). The displacement of the astrocytes also reveals the highly motile, smaller microglia, which are tightly associated with the astrocytic monolayer (green arrow). Scale bars: 20 µm; transmission light, time-stack imaging of astrocyte culture; 20xmagn.

The graph depicts the size of the holes in mm^2; n=6, the value represents the mean ± SEM;

B. Application of sub-lytic amounts of PLY reveals the outlining of cell nuclei (white arrows), probably induced by actin cytoskeleton contractions. Scale bar: 20 µm.

C. GFP-staining of the plasma membrane of primary astrocytes reveals the formation of membrane blebbings, visible continuously through the whole time stack. Scale bar: 10 µm.

D. Loosening of cell contacts and cell retraction induces holes or open spaces (red arrow). During the cellular displacement certain actin structures are retained (green arrow), while in general there occurs a decrease in actin density and actin structure complexity; visualization of actin with LifeAct-TagRFP, fluorescence imaging, 60xmagn.

Furthermore, an increase in cytoplasmic granules and stronger membranous fluctuations (wave-like) were observed, and the outline of the cellular nuclei became visible (Fig. 28 B).

Fluorescent labeling of actin with the help of LifeAct-TagRFP (red fluorescent protein) gave further insight into the cell displacement events. While the cells retracted, cell-cell contacts were loosened, and whereat in general a decrease in actin density and actin structure complexity was be observed, certain actin structures within the cells were retained (Fig. 28 D).

A very broad effect of toxin treatment was the so called membrane blebbing, a detachment of the plasma membrane from the cell body, which could easily be observed with probing the plasma membrane with GFP (Cellular Lights, PM-GFP) (Fig. 28 C).

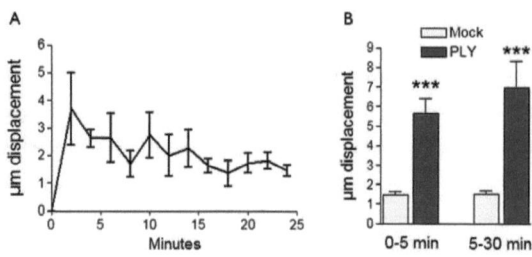

Fig. 29: *Phases of cell displacement.*
A. Cell displacement plot demonstrates a rapid initial phase (0-5 min) after treatment of astrocytes with 0.2 µg/ml PLY, and a slower, more moderate displacement thereafter.
B. Increase in cell displacement following PLY challenge. The increase in the second phase (5-30 min) is much lower than the displacement in the early phase before 5 minutes, indicated by the displacement difference; $p<0.001$, $n=10$ cells. The values represent the mean ± SEM.

Plotting of the cell-displacement (with the help of the displacement tracking assay, as described in the methods part) demonstrated a rapid initial phase (a few seconds to 5 minutes) with continuing but slower changes in the late period.

The shift in position within the first 5 minutes was very distinct (up to 6 μm from the original position), the changes in the following minutes were less severe and slower (only additional 1-2 μm shift in position).

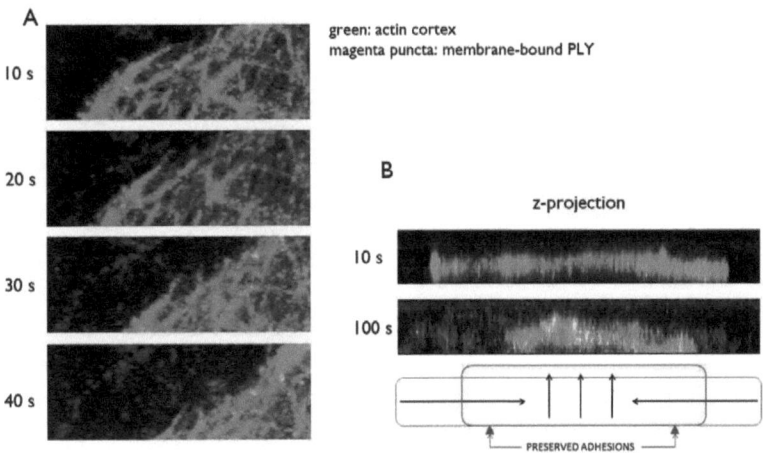

Fig. 30:   Actin cortex collapse.

Confocal high-resolution z-stack imaging of actin (visualized with LifeAct-TagRFP, green pseudo-coloration) and PLY (Atto488-tagged, magenta pseudo-coloration).

A. Within 40 seconds the actin cortex is severed from the plasma membrane and collapses in direction of the cell core. The membrane is indicated by the magenta puncta, comprising toxin aggregations on top of the cell.

B. Z-projection displays actin structure contraction in horizontal direction where biomechanical tensions are highest. The collapse happens in direction of the cell core, and thus leads to a vertical enlargement of the cell whereat cellular adhesions are preserved.

Fluorescent labeling of actin with RFP via transfection with the pLifeAct-TagRFP or the pTagRFP-actin vectors allowed as well a more detailed investigation of the cytoskeletal changes in the early phase after toxin application. High-resolution z-stack imaging revealed cell shrinkage (as seen before in transmission light live cell imaging), actin bundle deformation, and cortical actin bundling towards the cell core, while the Atto488-tagged toxin aggregated on the plasma membrane, and thus on the cell surface (Fig. 30 A). The decrease of actin density lead to mechanical imbalance, with the actin collapse occurring predominantly in horizontal direction where the tension powers are known to be highest [Pourati et al., 1998] (Fig. 30 B). The shrinkage caused a filopodia-like phenotype of the cells, which were keeping their adherence to the coverslip bottom (indicated in Fig. 33).

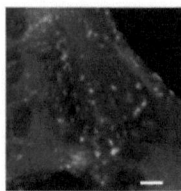

Mock

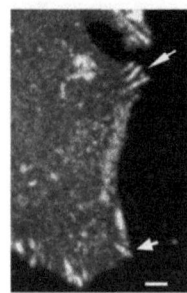

PLY 0.2 µg/ml

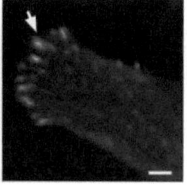

Fig. 31:   Development of focal adhesions.
During the cytoskeletal changes and cell shape remodeling induced by PLY challenge, distinct focal points (depicted in the mock-treated cells) expand to focal adhesions (FAs), which appear as longish aggregations of vinculin in immunocyto-chemistry staining. Scale bars: 10 µm;
confocal microscopy of fixed astrocytes after staining with a vinculin antibody; 63xmagn.

In the context of the observed cellular changes there also occurred an increase in the number and size of focal adhesions, which established from small, but defined focal points and developed upon PLY treatment in a time frame of about 30 minutes. FAs were visualized by vinculin immuno-staining in PFA-fixed cells (Fig. 31).

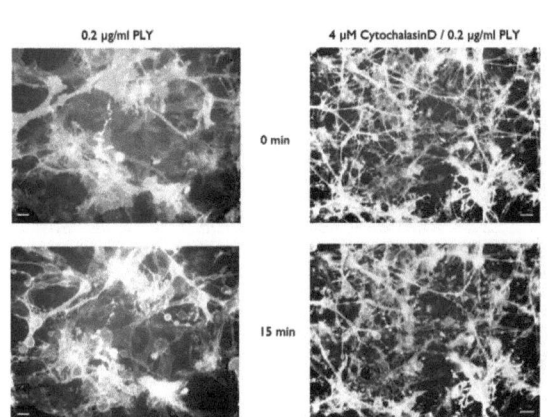

Fig. 32: Pre-treatment of astrocytes with cytochalasin D.

4 µM cytochalasin D lead to actin destruction in primary astrocytes, which abolishes displacement, and enhances the formation of membrane blebbing (right panel). Scale bars: 10 µm; epi-fluorescence imaging of astrocytes transduced with a Baculovirus-PM-GFP construct; 20xmagn.

Pre-treatment of the cells with cytochalasin D, a potent inhibitor of actin polymerization, and thus an actin cytoskeleton degrading chemical effector, potently inhibited most cell modifications induced by PLY. Movement and distraction of cells were completely abolished. Membrane blebbing, however, was not diminished by the mycotoxin, but enhanced (Fig. 32). The cells revealed in general a different phenotype, which derived from actin cytoskeleton degradation.

Visualization of actin, again with the help of the pTagRFP-actin vector allowed a closer look on the altered actin morphology induced by PLY treatment. Quiescent astrocytes showed a strong abundance of stressfibers, which spanned the cells (indicated by the green arrow in Fig. 33).

In a time frame of about 10 minutes after toxin application, those filamentous actin structures were degraded to a mesh-like phenotype of actin, probably comprising undefined actin aggregations or branched actin structures throughout the cell and around the nucleus. As mentioned before, the cells did not detach from the bottom of the culture slide, indicated by the establishment of filopodia-like structures (white arrows in Fig. 33).

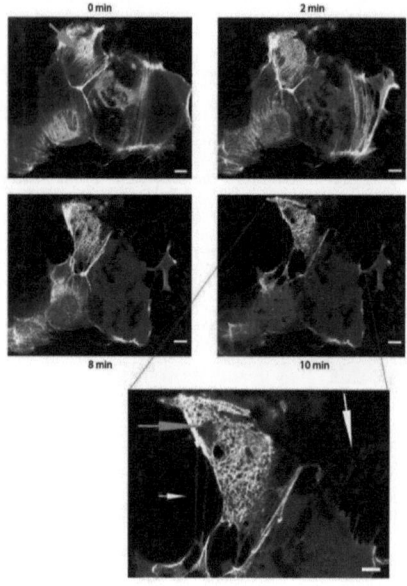

Fig. 33: Actin changes induced by PLY challenge.

0.2 µg/ml pneumolysin induces a change of actin morphology in astrocytes. Witin 10 minutes the filamentous actin is degraded to an actin mesh (both phenotypes indicated by green arrows). While the cells shrink, they keep attached to the bottom of the slide, which results in the establishment of filopodia-like structures (white arrows). Scale bar: 10 µm; Confocal fluorescence imaging after transfection with the pTagRFP-actin vector; 63xmagn.

To investigate the influence of calcium influx in the cell shape changes, the astrocytes were imaged (time lapse, transmission light imaging) in calcium free HEPES-buffer (see methods). The cells that were treated with 0.2 µg/ml PLY displayed a significantly higher displacement than the mock-treated cells. The displacement was comparable to the displacement of cells that were imaged in Leibovitz's medium.

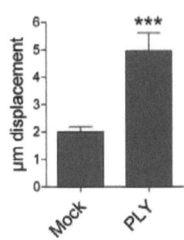

Fig. 34: Displacement of astrocytes in calcium-free buffer. Calcium-free conditions do not alter the range of cell displacement upon PLY treatment (0.2 µg/ml); ***p<0.001, n=30 cells from 4 experiments; the values represent the mean ± SEM.

2.4 Involvement of the small GTPases Rac1 and RhoA in the effects induced by pneumolysin

The dependence of the cytoskeletal changes on small GTPases was tested both biochemically, and in the cellular system. In the pull down assays (small GTPase activation assays) the activation levels of Rac1 and RhoA were compared to the levels in mock treated cells (primary astrocytes). Furthermore the levels of activated GTPases were normalized to the levels of total Rac and Rho in the cells.

Rac1 was upregulated after 5 minutes following PLY treatment, reaching a maximum of activation after 15 minutes. RhoA activation followed the same kinetics as Rac1 activation, but with generally lower amounts. Furthermore, RhoA activation stayed unchanged until 15 minutes after toxin application. After 30 minutes both GTPases reached a lower activation degree (Fig. 35, A). In the time frame of 0 to 5 minutes no activation could be observed.

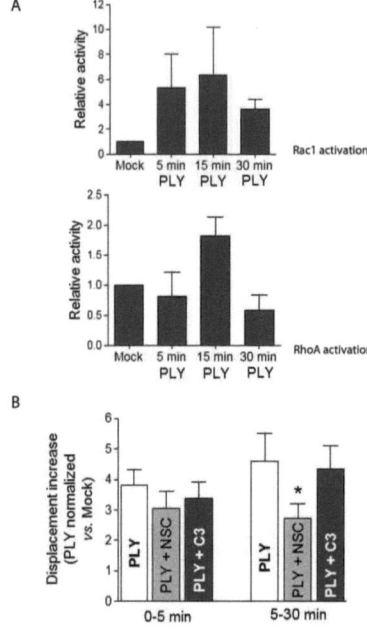

Fig. 35: Involvement of Rac and Rho in the cellular changes induced by PLY.

A. Pull-down analysis of activated Rac1 demonstrates an increase at 5 minutes following PLY challenge (0.2 μg/ml).
Pull-down analysis of activated RhoA demonstrates an increase at 15 minutes; n=3.

B. Increase in displacement (vs. mock) following a 0.2 μg/ml PLY challenge is not affected by NSC23766 (Rac1 inhibitor) and C^{3bot} transferase (Rho inhibitor) in the first 5 minutes. In the late window of displacement (5-30 min), the Rac1 inhibitor significantly reduces the displacement; *p< 0.05, n=30 cells from 4 experiments; all values represent the mean ± SEM.

For additional clarification a specific Rac1 inhibitor (NSC23677) and a general Rho inhibitor ($C3^{bot}$ transferase) were used. The displacement increase of inhibitor-treated cells after PLY challenge was compared to the untreated cells. Inhibition of Rho had no influence on the cell modifications induced by PLY, neither in the early phase (0-5 min), nor in the late phase (5-30 min). Rac1 in contrast was not changing PLY induced effects in the first 5 minutes, but significantly reduced displacement in the late phase, compared to untreated or $C3^{bot}$-treated cells (Fig. 35, B). Cellular uptake of $C3^{bot}$ was affirmed by a significantly different phenotype of astrocytes compared to untreated cells (data not shown).

2.5 Application of giant unilamellar vesicles (GUVs) for investigation of transmembranous PLY effects

The GUV system (giant unilamellar vesicles), which comprises a system of artificial membranes (vesicles consisting of a lipid-bilayer, with a diameter of about 10-30 µm), was used to investigate a possible direct interaction between pneumolysin and actin, while limiting the components to phospholipids, cholesterol, rhodamine-labeled actin (actin-TRITC), GFP-tagged toxin and Arp2/3 protein complex. The distribution of components, however, was inverted to the cellular system, with the cellular factors outside the GUVs and the toxin from within the GUVs. Loading of the vesicles with PLY was achieved by cold fusion (4°C) with a liposome carrier (schematic Fig. 36, A). Extra-vesicular PLY was inactivated with a cholesterol trap. Cells treated with the GUV solution showed no changes and no permeabilization, indicating successful depletion of PLY (data not shown).

Actin-rhodamine was resolved in general actin buffer, providing Arp2/3, ATP and Mg^{2+}, which allowed spontaneous nucleation and polymerization of actin, both in the suspension, and around the GUVs (Fig. 36, B, mock). Such aggregations were visible as small clusters of rhodamine after 30-60 minutes, and occurred also around the control GUVs, which contained no PLY. The presence of wild-type PLY, however, both accelerated, and increased the actin aggregation on the GUV surface, with accumulation of PLY-GFP co-localizing with the biggest actin-rhodamine clusters (Fig. 36, B, WT-PLY).

The non pore-forming mutant Δ6-GFP itself aggregated partially, but was not able to initiate actin clustering in a similar magnitude as the wild-type toxin (Fig. 36, B, d6-PLY).

The stabilization of actin was measured by comparison of the amount of pixel with a certain intensity. All values were normalized to the background comprised by rhodamine fluorescence. The GUVs, which contained the WT-PLY, showed the highest number of pixel with an intensity above the background.

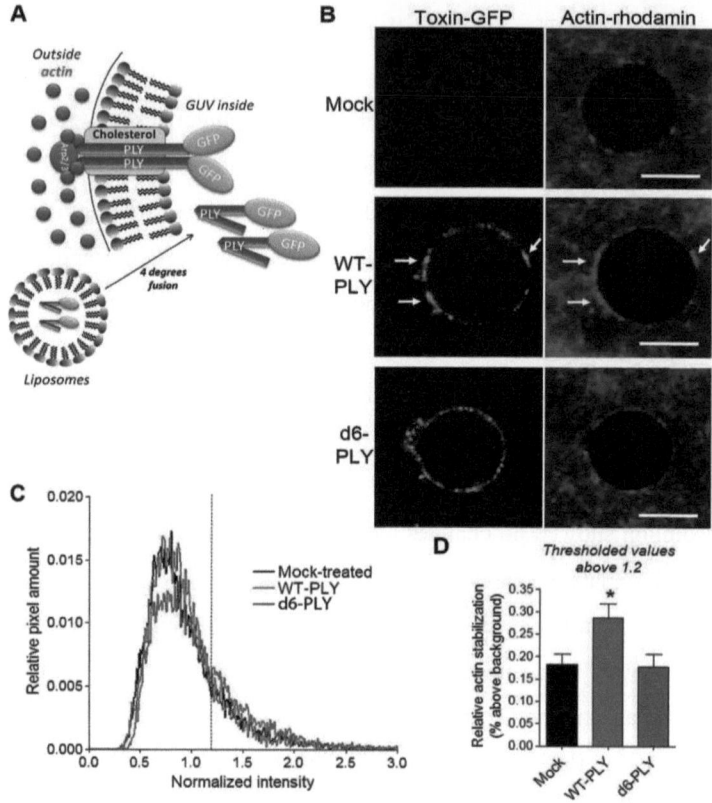

Fig. 36: GUV approach.
A. *Schematic presentation of the GUV experimental approach. The toxin and the actin together with the Arp2/3 complex localize on both sides of the lipid bilayer in a manner corresponding to the cell membrane.* ***B.*** *Aggregation of actin-rhodamine on the surface of GUVs in close proximity with the PLY-GFP clusters of wild-type PLY (WT-PLY; arrows), but not in the mock sample or the non-pore forming delta6-PLY (d6-PLY) GUVs. Scale bar: 10 μm* ***C.*** *Cumulative histogram of the pixel intensity distribution along the GUV surface normalized to the average actin-rhodamine intensity outside the GUV. The histogram represents cumulative data of 15-20 GUVs of one representative experiment. The experiment was replicated 3 times.* ***D.*** *Increased actin aggregation (as measured by the increased rhodamine intensity above the actin background) in the WT-PLY GUVs; *p <0.05, values represent the mean ± SEM, n =15-20 GUVs of one representative experiment, replicated 3 times.*

The actin accumulation around the GUVs, which contained the Δ6 mutant, was comparable to the clusters that occurred around the un-loaded GUVs, and which were due to spontaneous actin nucleation in the suspension (measurement of pixel intensity depicted in Fig. 36C, D).

WT-PLY was able to stick through the lipid bilayer and aggregate actin. Arp2/3 was present in the system. It nucleated actin in absence of VCA, however, only in combination with WT-PLY (Fig. 36, B).

2.6 Investigation of PLY-actin interaction

2.6.1 Spin-down assays

The capacity of pneumolysin to directly interact with actin and influence the actin status was investigated mainly with the help of biochemical approaches, both cell-based, and purely biochemical, using human platelet actin and isolated ABPs (actin binding proteins).

When cells were treated with PLY (0.2 µg/ml) and the cytoskeleton subsequently isolated and spun down by ultracentrifugation, the toxin could be detected in the pellet fraction (Fig. 37). At a speed of 150.000 x g, a 53kD protein (such as PLY) only settles down in the pellet when bound to a protein with higher molecular weight. Actin filaments and tubulin filaments have a weight of a multiple of 42 kD for actin and 55 kD for tubulin, respectively.

Upon degradation of the actin or tubulin cytoskeleton, or both, there was less toxin found in the pellet fraction, meaning less toxin settling down in the pellet during ultracentrifugation. With less extend of cytoskeleton structures to bind to, the amount of detectable toxin became smaller

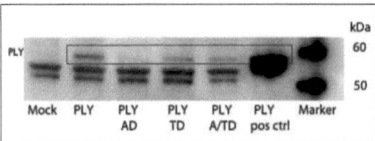

Fig. 37: **Cytoskeleton Isolation Assay.**
PLY co-sediments with the cell cytoskeleton during ultracentrifugation. Degradation of actin (AD) or tubulin (TD) or both (A/TD) decreases the toxin amount in the pellet. (Western Blot, anti-PLY staining)

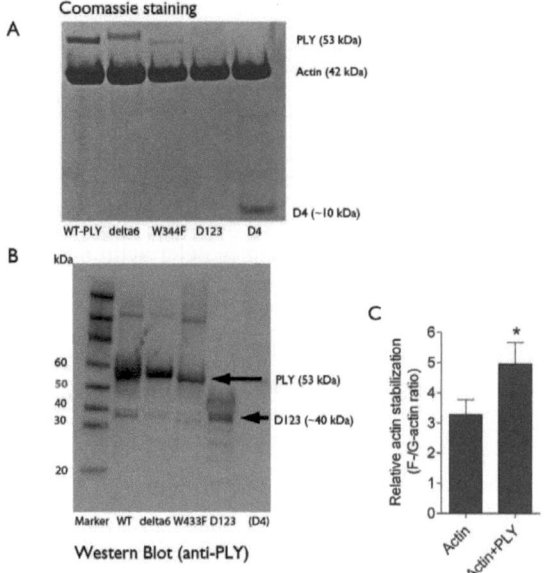

Fig. 38: **Actin binding and stabilization capacity of PLY.**
PLY co-settles with isolated actin during ultracentrifugation. **A.** Coomassie staining of the pellet fraction containing F-actin that co-precipitates with WT-PLY and weakly co-precipitates with Δ6-PLY, W433F-PLY and D4-PLY. D123-PLY overlaps with actin. **B.** Western Blot analysis of the co-precipitation of the D123-PLY confirms its presence in the pellet at ~10 kD. **C.** PLY significantly stabilizes actin (F-/G-actin ratio from the precipitation assays); *$p > 0.05$, $n = 5$, values represent mean ± SEM.

To clarify the capacity of PLY to bind to actin filaments (F-actin), an actin binding spin down assay was used, which reveals a possible co-precipitation of a certain protein together with actin. Isolated human platelet actin was incubated together with WT-PLY, the mutants (Δ6 and W433F), and the fragments (D123, D4), and subsequently ultracentrifuged. The supernatant fraction and the pellet fraction were electrophoresed and the gels stained with Coomassie. The proteins were blotted on PVDF membranes, and probed with a PLY antibody. The spin down assay revealed that every variant of pneumolysin was able to settle down together with actin in the pellet, indicating that every variant had the capacity to bind the toxin. The wild type toxin, however, bound actin with the highest affinity. All variants were detectable with Coomassie staining, except the D123 fragment, which has about the same molecular weight as actin, and thus was hidden behind the actin band (see Fig. 38, A). Domain 4 in contrast could not be detected with the PLY antibody, as latter is specific for binding to the domain 123 fragment (see Fig. 38, B).

Densitometric analysis of the actin bands in the supernatant and pellet fractions revealed that incubation with PLY slightly shifted the G-actin/F-actin ratio in direction of fibrous actin, which indicates a stabilization of actin (Fig. 38, C).

2.6.2 Actin-pyrene polymerization assays

To further investigate a possible stabilization/nucleation capacity of PLY, an actin-pyrene polymerization assay was used. Pyrene develops fluorescence as soon as it polymerizes. Actin is only able to polymerize in the presence of ATP and Mg^{2+}, both of which were provided in the polymerization buffer. The pyrene-actin in general actin buffer was monitored for 5 minutes and comprised background fluorescence. Addition of ATP and Mg^{2+} to the actin suspension lead to a slight increase of pyrene fluorescence. This phenomenon also occurred when PLY was added together with the polymerization buffer, whereat the fluorescence intensity was higher with PLY than with the buffer alone. The increase induced by PLY occurred in a dose-dependent manner (see Fig. 39). PBS, which comprises the solvent of pneumolysin, was not able to influence the fluorescence level.

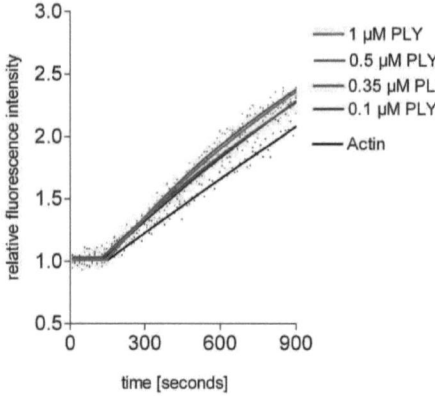

Fig. 39: *Actin stabilization capacity of PLY.*

Dose-dependent stabilization of actin-pyrene (3 µM) by PLY (buffer supplemented with Mg^{2+} and ATP) as judged by measurement of fluorescence intensity increase. 100 nM of PLY are sufficient to increase the amount of polymerized actin.

2.6.3 Far Western Blotting

The capacity of a protein to bind another protein, even if this one is partially degraded, can be easily detected with overlay blotting (also called Far Western Blotting [Hall, 2004]). Human platelet actin and RhoA recombinant protein (comprising the control) were electrophoresed, applied to Western blotting and the membranes were incubated together with (= PLY overlay) or without PLY. The subsequent staining with a PLY antibody displayed, that PLY was able to strongly bind actin, even in a partially degraded conformation. In contrast, there was no binding of PLY to the RhoA control protein (Fig. 40).

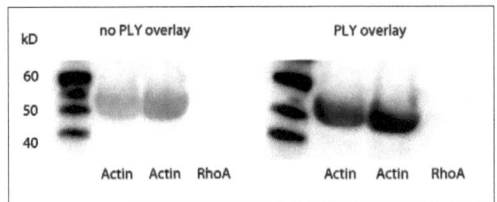

Fig. 40: Overlay blot.
Overlay and detection of PLY after electrophoresis and blotting of actin and RhoA. PLY shows a strong affinity to actin (vs. RhoA) even in a denatured conformation of the protein.

2.6.4 Enzyme-linked sorbent assays (ELSAs)

G-actin binding was tested with the help of an enzyme-linked sorbent assay (ELSA). WT-PLY was linked to the bottom of a maxisorp plate and actin or FCS (control) was added.

The bound actin was detected with the help of a primary actin antibody and a HRP-conjugated secondary antibody, which catalyzes, dependent on the amount of bound antibody, a calorimetric change in the TMB substrate (Fig. 20 displays a schematic of the assay).

Only actin, and none of the proteins of the serum (FCS) were able to interact with PLY, which led to a higher absorption value in the test (Fig. 41).

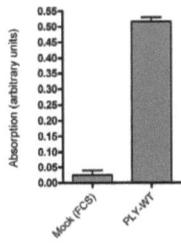

Fig. 41: Affinity of PLY to actin.
Enzyme-linked sorbent assay.
G-actin binding of PLY-WT in comparison to FCS; n=3, values represent the mean ± SEM.

2.6.5 Immunocytochemistry

Confocal microscopy of pneumolysin-treated and subsequently PFA-fixed astrocytes indicated that PLY had a certain affinity to actin also in the cellular system. The actin cytoskeleton was visualized with the help of phalloidin-Alexa555. The cells were fixed at very early time points (in the range of 10 to 60 seconds) after PLY-Atto488 challenge, right in the state of membrane-binding, before the toxin was able to induce disturbance of the actin cytoskeleton. PLY was found to localize on the membrane in close vicinity to the actin cortex structures below the PM, decorating the actin bundles from this membrane-bound state (Fig. 42). The picture in the middle (lower panel) depicts the toxin aggregates on the membrane. The merge ("overlay") shows the intimate proximity of both proteins.

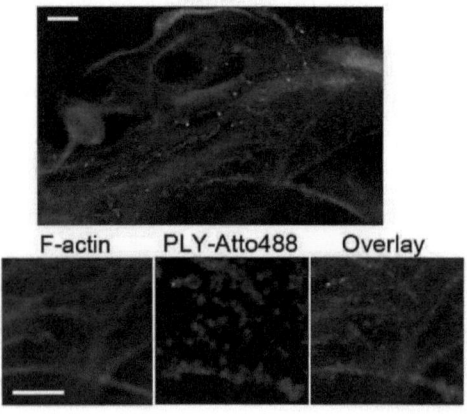

Fig. 42:   *Actin bundle decoration by PLY.*

Imaging of astrocytes demonstrates a co-localization between F-actin (stained with phalloidin-Alexa555, green pseudo-coloration) and PLY (Atto488-tagged, magenta pseudo-coloration) 60 s after toxin exposure (0.2 µg/ml) on the top of the cell. Scale bars: 10 µm.

2.7 Investigation of PLY-Arp interaction

2.7.1 Enzyme-linked sorbent assays (ELSAs)

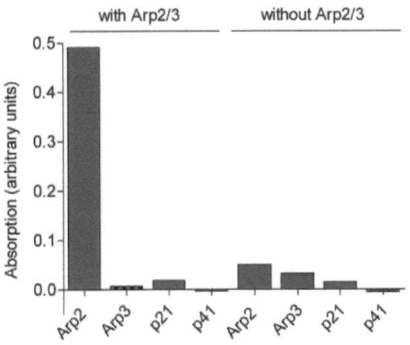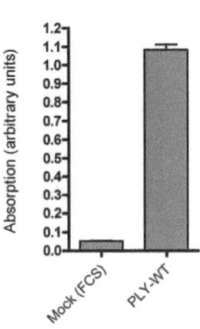

Fig. 43: Binding of PLY to the Arp complex and to Arp2.
An ELSA (Enzyme-linked sorbent) assay demonstrates an increased binding affinity of PLY to the Arp2 component of the Arp2/3 complex, and compares the binding capacity of PLY to Arp2 in comparison to FCS; n=3, values represent the mean ± SEM.

Arp2/3 is a heptameric protein complex. Four components of the complex, namely Arp2, Arp3, p21 and p41 were tested for interaction with pneumolysin with the help of an enzyme-linked sorbent assay (ELSA). WT-PLY was linked to the bottom of a maxisorp plate, and the binding of Arp components, or Arp2 vs. FCS, respectively was judged by measurement of absorption (as described above).
Only Arp2, and not Arp3 or p21/p41 of the Arp complex showed the capacity to significantly interact with WT-PLY (Fig. 43). Serum proteins (FCS) hat no capacity to interact with PLY.

2.7.2 Immunocytochemistry

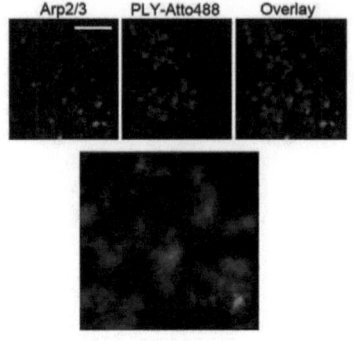

Fig. 44: ***Co-localization of PLY and Arp2.***
Shortly after toxin exposure (<1min), PLY-Atto488 (magenta pseudo-coloration) clusters in immediate vicinity or directly co-localizing with Arp2 structures (green pseudo-coloration); confocal z-stack imaging. Scale bar: 10 µm.

Similar to the investigation of PLY-actin interaction, PLY was tested for its capacity to co-localize with Arp2. Therefore, astrocytes were treated with PLY-Atto488 and subsequently fixed with paraformaldehyde (PFA). Arp2 of the Arp2/3 complex was visualized by tagging with an Arp2-specific antibody. The toxin was found to cluster rapidly (within the timeframe of 1 minute) and very closely and partially overlapping with Arp2, which is depicted in the overlay of Fig. 44.

2.7.3 Actin-pyrene polymerization assays

Arp2/3 is a potent nucleator of actin, but requires activation by the VCA domain of WASP family proteins. The development of fluorescence increase in the presence of VCA is shown in Fig. 45. The fluorescence developed rapidly, reaching a higher maximal value in comparison to the activation of Arp by PLY.

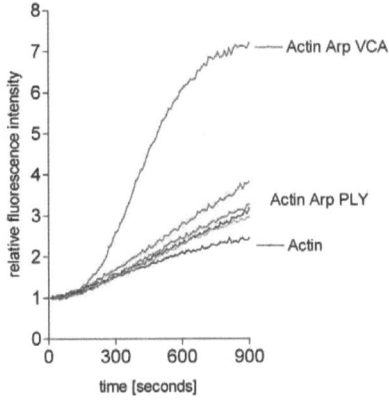

Fig. 45: *Arp2/3-mediated actin polymerization.* Comparison of Arp2/3 activation by PLY (100 nM-1µM) and the cellular Arp2/3 activator VCA domain (50 nM).

In actin-pyrene polymerization assays, Arp2/3 protein complex alone slightly increased actin polymerization and thus fluorescence (Fig. 46 A, Actin+Arp). Arp in combination with PLY enhanced clearly the pyrene fluorescence (Fig. 46 A, Actin+Arp+PLY), similarly to the experiments without Arp (compare to Fig. 39), in a dose-dependent manner (Fig. 46 B).

Thus, Fig. 46 A shows that PLY alone was able to only slightly stabilize actin, but to potentially activate Arp2/3, which lead to an increase in actin-pyrene fluorescence.

In the presence of VCA domain protein, the fluorescence showed the fastest and strongest increase, which is typical for actin nucleation by VCA-activated Arp. In presence of VCA and PLY the fluorescence reached the same maximum, but even faster, with the fluorescence curve showing a stronger gradient than the VCA curve (Fig. 47).

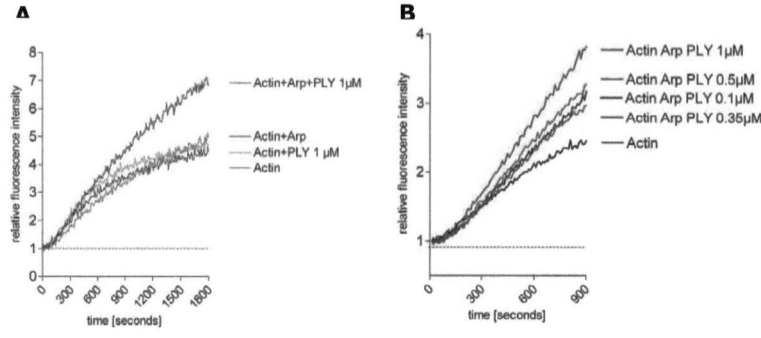

Fig. 46: **Arp activation capacity of PLY.**
Comparison of actin polymerization by PLY, non-activated Arp2/3 (50 nM), and PLY-activated Arp2/3. Arp2/3 activation by PLY occurs in a dose-dependent manner, starting at low toxin concentrations (100 nM-1 µM PLY) already.

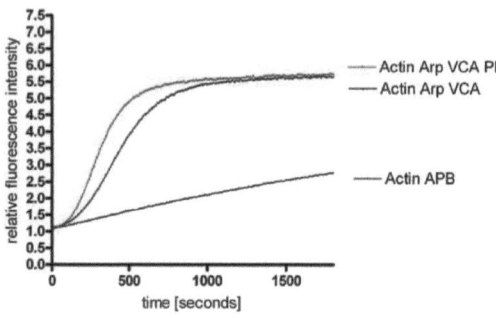

Fig. 47: **Additive effects of PLY in the activation of Arp2/3 by VCA.**
VCA (50 nM) and PLY (1 µM) work in a synergistic manner to activate Arp2/3 (50 nM), which results in an acceleration of actin nucleation.

2.7.4 Imaging of actin-rhodamine (TRITC-tagged)

Coverslip chamberslides were coated with streptavidin, and pre-formed actin filaments, consisting of non-muscle actin, actin-rhodamine (actin-TRITC) and actin-biotin were applied to fluorescence imaging. Addition of 1 µM pneumolysin to the actin filaments did not change their morphology. The addition of Arp2/3 protein complex resulted in the Arp-specific branched actin structures. Here it was not relevant if Arp was activated by VCA or by PLY.

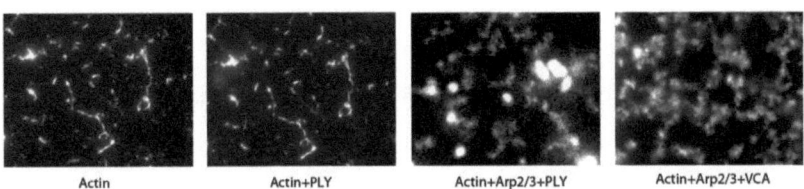

Fig. 48:   Fluorescent imaging of actin-rhodamine.
PLY (1 µM) has the capacity to activate Arp2/3 (60 nM) in a similar mode as VCA (60 nM). The resulting actin structures (Actin+Arp2/3+PLY and Actin+Arp2/3+VCA; 200 nM actin) reveal a similar morphology; epi-fluorescent microscopy, 60x magn.

2.8 Investigation of structural characteristics of PLY-actin and PLY-Arp2 interaction

2.8.1 Enzyme-linked sorbent assays

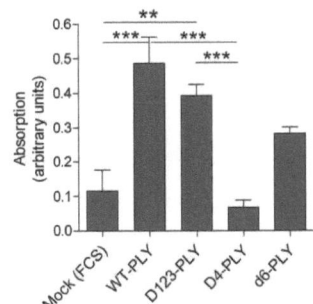

Fig. 49:   Structural properties of PLY interaction with G-actin.

*An ELSA (enzyme-linked sorbent) assay demonstrates an increased binding affinity of WT-PLY and the 1-3 domains (D123-PLY) towards G-actin and reduced affinity of the Δ6-PLY mutant and the binding domain 4; **$p<0.01$ ***$p<0.001$, n=6, values represent mean ± SEM.*

Pneumolysin is a 4-domain protein, with its binding motif in domain 4 and domain 3 as unfolding domain. The availability of variants and fragments of the toxin allows a more detailed investigation of interaction of PLY with other proteins or the membrane.

Application of ELSAs (enzyme-linked sorbent assays) clearly showed that only the wild-type toxin and the D123 fragment had the capacity to strongly interact with actin and Arp, whereas the binding domain (D4) and the Δ6 mutant, which cannot unfold properly in the membrane-bound state, revealed less affinity to actin and Arp (Fig. 49, Fig. 50).

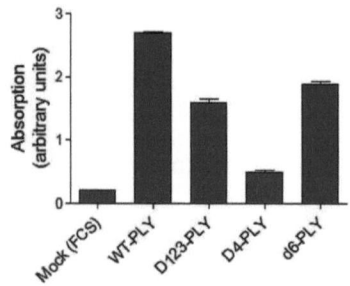

Fig. 50: *Structural properties of PLY-Arp2 interaction.*

An ELSA (enzyme-linked sorbent) assay demonstrates an increased binding affinity of WT-PLY and the 1-3 domains (D123-PLY) towards Arp2 and reduced affinity of the Δ6-PLY mutant and the binding domain 4; n=3, values represent the mean ± SEM.

2.8.2 Actin-pyrene polymerization assays

In order to gain a better insight into the structural prerequisites of interaction and binding between PLY and actin/Arp, single peptide sequences of the unfolding domain (D3) were used in actin-pyrene polymerization assays to check their possible influence on Arp-activation by PLY. The peptides have the following positions in D3:

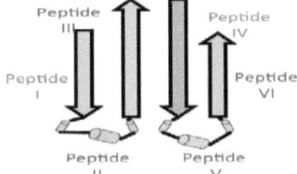

Fig. 51: ***Schematic of the unfolding domain of PLY.***
Domain 3 of PLY consists of 6 peptides, which refold into β hairpins that penetrate into the membrane.

Adding the single peptides (I-VI) to actin, in combination with Arp, actin polymerization buffer and PLY, revealed no significant change in fluorescence increase. Only peptide II increased the fluorescence to a slightly higher maximal value. The effects of the peptides are depicted in Fig. 52.

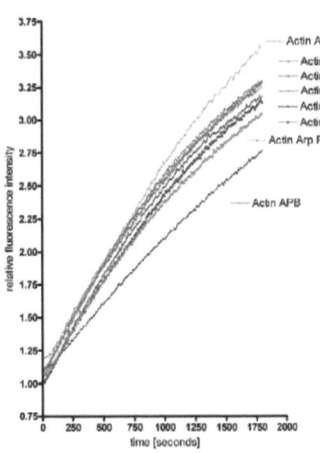

Fig. 52: ***Investigation of structural properties of PLY-Arp interaction.***
Simultaneous incubation of Arp2/3 (50 nM) with PLY (1 μM) and the single peptides (10 μM) of the unfolding domain 3 does not alter actin polymerization kinetics.

2.8.3 FRET (Fluorescence resonance energy transfer)

FRET experiments uncovered a significant fluorescence energy transfer between a PLY-Atto488/Phalloidin-Alexa555 FRET pair (Fig. 53). An increase in phalloidin concentration quenched the PLY-Atto488 fluorescence in a concentration-dependent manner.

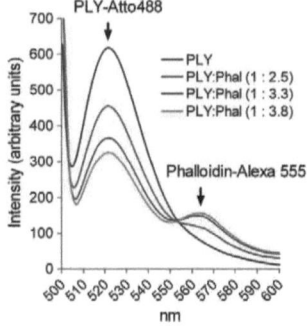

Fig. 53: **FRET measurement between PLY and phalloidin.**
Spectrophotometric analysis of the decrease in PLY-Atto488 fluorescence after addition of phalloidin-Alexa555, which is indicative of FRET and confirms an interaction between PLY and phalloidin.

2.9 Influence of the Arp2/3 complex on pore formation - application of erythrocyte ghosts

In order to examine the role of the PLY/Arp interaction on the capability of PLY to form lytic pores an erythrocyte system was used. This system avoided the interference of cell trafficking phenomena on the toxin's lytic capacity, as it is devoid of endo-/exocytotic events. The ghosts were depleted of ATP, and all other cytosolic components) and resealed after lytic emptying.

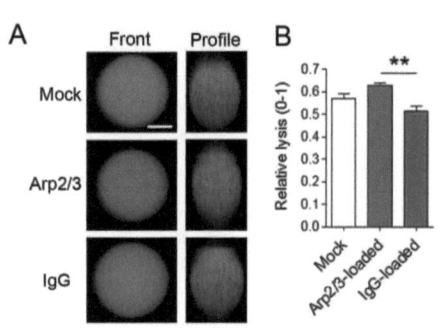

Fig. 54: **Influence of Arp2/3 on the lytic capacity of PLY.**
A. Confocal imaging of calcein-loaded resealed ghosts and z-profile reconstruction confirm the retained erythrocyte-like disc shape, proper resealing and equivalent size. Scale bar: 2 µm.
B. Relative lytic capacity of 0.2 µg/ml PLY on calcein-loaded resealed ghosts, containing either the Arp2/3 complex, nor-specific IgG or nothing. The enrichment with Arp2/3 enhanced the lytic capacity of PLY vs. the IgG-control; **p<0.01. Values represent mean ± SEM.

With the help of confocal imaging it was made sure that the resealed ghosts had retained the erythrocyte-like disc shape (Fig. 54 A). The different loading of the erythrocytes did not influence the shape. The experiments revealed that the lytic capacity of PLY in calcein-loaded, Arp2/3-enriched ghosts was significantly higher than in IgG-enriched or mock ghosts.

3 DISCUSSION

Bacterial meningitis is defined as an inflammation of the protecting membranes of brain and spinal cord, the meninges, with impact on the pia, the arachnoid and subarachnoid space, and remains the most common form of CNS infection in humans. It is associated with a severe clinical outcome, high case fatality rates, and high incidence rates of remaining neurological deficits (reviewed in [Kim, 2003]). It was supposed that the disease-related symptoms, such as brain edema (acute) and learning disabilities (chronic), derive from modulations induced by the bacteria and their effector proteins in the brain tissue, and additionally from the destructive force of immune responses of the host itself (reviewed in [Gerber and Nau, 2010]).

In order to investigate the molecular mechanisms of cell modifications in the context of pneumococcal meningitis, the present study used primary rodent astrocytes (mainly deriving from mouse brain), and disease-relevant concentrations of pneumolysin, which is a key virulence factor of *S. pneumoniae*, as its presence, amongst other various effects, aggravates the clinical course of meningitis [Hirst et al., 2008; Hirst et al., 2004].

Structural studies, which normally use up to 100-fold higher concentrations of PLY ($\sim$ 10 µg/ml), focus on the pore formation and lytic capacities of the toxin, inducing destruction of cells (hemolysis) or liposomes [Berry et al., 1995; Bonev et al., 2000; Gilbert et al., 1999; Tilley et al., 2005]. These concentrations, however, have no relevance in disease-related studies ($\leq$ 0.18 µg/ml pneumolysin can be detected in the liquor of meningitis patients [Spreer et al., 2003]).

Meningitis is a disease mainly affecting newborns, children younger than 5 years of age (indicated in Fig. 3), and immune-suppressed patients. Thus, in order to investigate neuronal cells in a disease-relevant state of development, newborn mice (p 3-5) were chosen for the isolation of astrocytes from brain tissue.

The astrocytic nature of those cells was confirmed by fluorescent staining with a specific GFAP antibody. The glial fibrillary acidic protein (GFAP) is the principal intermediate filament in mature astrocytes and is widely used as a marker for the identification of cells in nervous tissue, and especially for the immunocytochemical/and -histochemical detection of astrocytes (reviewed in [Eng et al., 2000; Sofroniew and Vinters, 2010]).

Although GFAP cannot be considered as an absolute marker, as it is present in most, but not all astrocytes, it could be detected in a wide range of the cultured cells (Fig. 25). The presence of neurons in the cultures could be generally excluded, as culturing of those cells requires special and well-adapted media, whereas the astrocytes are able to grow in D-MEM, supplemented with serum (FCS). However, the investigation was set to be limited to astrocytes, and thus the presence of large numbers of other brain tissue components, such as fibroblasts, had to be excluded.

The star-shaped phenotype, which was already realized by Ramon y Cajal and del Rio Hortega in 1909, occurred also in the fluorescent GFAP staining (Fig. 25). They described two different subtypes of those glial cells, the protoplasmic astrocytes of the gray matter, which show a morphology of several stem branches that give rise to many finely branching processes, and the fibrous astrocytes of the white matter, which exhibit a morphology of many long fiber-like processes [Ramon y Cajal, 1909].

In a previous study the liquor of patients suffering from acute pneumococcal meningitis was tested for its toxin load, and CSF (cerebrospinal fluid) concentrations of PLY were found to range between 0.001 and 0.2 µg/ml [Spreer et al., 2003]. As this study used *in vitro* experiments (cell-based or biochemical), and not *in vivo* conditions, applicable concentrations of PLY had to be determined. Those concentrations were extrapolated to be disease-relevant, as well as they were supposed to reveal a sub-lytic behavior to the cells, as the aim was to investigate the effects of PLY in the absence of cell death.

Challenging primary astrocytes with concentrations of PLY between 0.05 and 0.5 µg/ml revealed cell permeabilization of 5% at 0.05 µg/ml, and 50% at 0.5 µg/ml, respectively, as judged by propidium iodide (PI) staining. At 0.1 µg/ml (disease-relevant concentration) only about 10% of cells were lysed (Fig. 26). However, the amount was sufficient to remodel the cell shape and to induce displacement of astrocytes. Many of the PI-positive cells were identified as microglia, which could be distinguished from the astrocytes by cell shape and size, and as they localize on top of the astrocytic cell layer [Squire, 2003]. These results enabled to define concentrations of 0.1 - 0.2 µg/ml of PLY as sub-lytic to non-lytic, comparable with the CSF concentrations of toxin in the liquor of patients [Förtsch et al., 2011]. Thus, under "sub-lytic" we defined the conditions, in which only a small portion of cells would lose their integrity (~ 5%), while the cytoskeleton cell shape effects would be visible in most of the surviving cells. Dependent on the toxin batch, each of which contained different amounts of active PLY monomers, the concentration was adjusted to correspond to an identical level of lytic units.

The challenge of astrocytes with 0.2 µg/ml pneumolysin led to very distinct cell shape changes and cell displacement. The contraction of the cell bodies induced the disruption of the cell monolayer, established by astrocytes when they are cultivated *in vitro* (Fig. 28). The displacement could be divided into an early phase with strong changes and a late phase with less severe changes, as shown in Fig. 29. The movement of the PLY-treated astrocytes could be distinguished clearly from the slight cellular fluctuations of quiescent cells, and thus clearly be assigned to as effects induced by PLY. Plasma membrane staining with GFP revealed that the cells kept attached to the bottom during their displacement (Fig. 32, Fig. 33), confirming that they were alive and that the shape changes were not due to cell death. Fluorescent staining of vinculin (a protein-component of focal adhesions, FAs) uncovered an increase in number and size of FAs upon toxin treatment (Fig. 31), which is conform to a model that describes the relation between adhesion and motility [Murphy-Ullrich, 2001]. The presence of defined focal points in astrocytes under untreated conditions indicates a capacity of motility. The displacement and retraction effects, followed by a less motile phase, accompanied by increased adhesion as witnessed by the progression from focal points to focal adhesions, fit very well with the concept of a transition from intermediate to strong adhesion separated by a mobile step (as discussed in [Förtsch et al., 2011]). The cell shape changes were accompanied by the appearance of nuclear outlining, which uncovered the roundish structure of the cell nuclei, which are normally not clearly visible during bright field, light microscopy (depicted in Fig. 28). This phenomenon is likely to be linked to the cytoskeletal changes occurring during displacement, which include cellular tensions, and which are also connected to actin remodeling (indicated in Fig. 9). Normally most of the astrocytes are quite static cells, which is a prerequisite for their supportive function in the brain (see the introduction). Thus the occurrence of strong cell displacement induced by PLY, although not untypical, is clearly not common for the astrocytes.

Each of the described effects, shape-changes, displacement via active movement (as recently reviewed by Anne Ridley [Ridley, 2011]), and establishment of focal adhesions are exclusively dependent on the actin cytoskeleton [Murphy-Ullrich, 2001], and the capacity to rapidly move is a defining feature of animal cells [Ridley et al., 2003]. Actin filaments are the basic units of cell locomotion. Disassembly of those filamentous structures by the use of cytochalasin D clearly showed a dependency of the PLY-induced effects on a functioning actin treadmilling apparatus, as the lack of actin structures abolished virtually all cell shape modifications, except the formation of membrane blebbings (Fig. 28 C), which in contrast were more abundant in the presence of CytoD (Fig. 32). Cytochalasins are fungal metabolites (CytoD is deriving from *Zygosporum mansonii*) and inhibit actin polymerization by high-affinity binding to the growing ends of actin filaments, preventing the addition of monomers [Casella et al., 1981]. This explains the inhibition of actin filament generation, which also prevents the rebuilding of the actin cortex below the membrane, and thus the retraction of the blebs. The altered phenotype of the astrocytes after cytochalasin pretreatment, and the lack of motility (Fig. 32) derive from the cytoskeletal destruction as well.

Membrane blebbings are classically a feature of apoptosis (programmed cell death), and occur when the plasma membrane detaches from the actin cortex, which is anchored directly to the plasma membrane (PM). Increasing osmotic and onkotic pressure leads to a protrusion of the PM. In case of apoptotic membrane blebbings, the cortex rebuilds at sites of remaining actin structures, leading to a retraction of the PM after a short time interval (about 30 seconds). Furthermore, the apoptotic blebbings occur late in the progress of the cell death, normally hours after the initial pathogenic insult (reviewed in [Charras, 2008]). In contrast to these apoptotic blebbings, the ones observed after toxin-treatment of astrocytes did not show any signs of retraction, but the protrusions remained present over the whole time course of the experimental procedure (Fig. 32). This can only be explained with an ability of PLY to inhibit the rearrangement of actin below the plasma membrane, which can be achieved by a stabilization or severing capacity of the toxin. Live cell experiments with LifeAct-TagRFP-stained actin and PLY-Atto488 indicated a similar action of pneumolysin, as the actin cortex showed a rapid detachment of the membrane and retraction in direction of the cell core (Fig. 30 A). The position of the PM was witnessed by the fluorescent punctate clusters, which were aggregating on the top of the cells and maintaining their position even when the actin had detached. These dots represented most probably toxin aggregates, which seemed to promote severing of the actin cortex, following membrane binding and pore formation.

These rapid events after toxin application reveal striking similarity in the mode of action between pneumolysin and the cellular actin effector gelsolin. Gelsolin was identified as one of the most potent actin filament severing proteins. Severing is mediated by the weakening of non-covalent bonds between actin molecules within one filament in order to break the filament in two. After severing, gelsolin keeps attached to the barbed end of the filament as a cap, which results in short filaments that can neither reanneal, nor be further elongated at their barbed ends. The generation of such short filaments can, on the one hand, result in the disassembly of the actin cytoskeleton, although it can also have a constructive effect, as it increases the number of filaments with polymerization-competent ends, as soon as gelsolin dissociates from the elongation sites (the properties of gelsolin are reviewed in [Sun et al., 1999]). One could speculate that PLY severs actin filaments from the PM, which induces the collapse of the cortex, and thus increases the number of barbed ends, which in turn provokes actin polymerization, possibly even at more or less adjacent sites, as specified further down. This is consistent with the transformation of actin fibers to an actin mesh, as shown in Fig. 33.

Interestingly, the detachment of actin from the PM occurred in a horizontal direction, meaning laterally in direction of the cell core, whereat the actin cortex on top of the cells maintained its attachment (Fig. 30 B). This phenomenon can be easily explained with biomechanics, as it is known that the highest impact of tensions occurs at the sides of the cell, whereas the impact of vertical tensions is much lower. This might lead to an instability of the whole actin cytoskeleton scaffold as soon as PLY exerts its interfering actions at places where dragging forces are highest [Pourati et al., 1998].

Although LifeAct is very suitable for the visualization of actin structures in eukaryotic cells, it comprises a relatively new approach. Thus, it was necessary to confirm the results that were achieved with this system with a more reliable and established approach. Transformation of astrocytes with the pTagRFP-actin vector revealed similar actin changes in the astrocytes treated with PLY, compared to the LifeAct transfected cells, namely a decrease in actin network density and a general conversion of actin structures (Fig. 33).

Calcium influx is a potent activator of GTPases, such as RhoA [Rao et al., 2001] and Rac [Price et al., 2003], and is often involved in the induction of cellular shape changes or cell death. PLY can induce early calcium influx [Stringaris et al., 2002], which might lead to the modulation of actin dynamics [Evans and Falke, 2007]. Therefore, the astrocytes were challenged with PLY in calcium-free buffer. In these experiments the concentrations of the toxin had to be adjusted due to earlier findings that PLY had an increased lytic capacity under calcium-free conditions [Wippel et al., 2011]. Even in the absence of extracellular calcium, pneumolysin still retained its capacity to produce rapid cell shape changes in astrocytes (Fig. 34), proving the independence of the observed effects on calcium influx.

Cell modifications by bacterial toxins are often related to changes of the actin cytoskeleton, whereat the modulation results from direct or indirect modification of the cytoskeletal apparatus through covalent or non-covalent mechanisms (reviewed in [Barbieri et al., 2002]). One common non-covalent, indirect mechanism of those toxins is the modulation of small GTPases.

Small GTPases are molecular switches that cycle between an on-state when bound to ATP and an off-state when bound to ADP. Subcellular signaling events, mainly arriving from the exterior of the cell, lead to activation of GEFs (guanine nucleotide exchange factors) or GAPs (GTPase activating proteins) and thus directly influence the status of the switch by controlling the nucleotide load (for further review see [Ridley, 2001]. The small GTPAses Rac1 and RhoA are the two main regulators of the actin cytoskeleton [Aspenstrom et al., 2004], being responsible for pushing cells forward by establishment of lamellipodia at the front and contraction of stress fibers, which reach throughout the cell body. The two proteins have many more functions in the context of cell behavior, and the findings that sub-lytic concentrations of pneumolysin induced simultaneously an upregulation of Rac and Rho, and cytoskeletal rearrangement, were the cornerstone of this work [Iliev et al., 2007].

However, neuroblastoma cells, applied in the initial phase of studying PLY effects, comprise a tumor cell line, and thus results can vary or be less relevant to the *in vivo* situation. In order to better understand the effects of pneumolysin in the physiological context of meningitis, and to gain more stable and reliable results, this study used primary astrocytes. Astrocytic cultures, however, are mixed cultures with microglial cells, another cell type of brain tissue. Microglia are immune-active cells, originating from the monocyte/macrophage line and sharing similarities with the macrophages, the Kupffer cells in the liver and the mesangial cells of the kidney. Microglia use the astrocyte syncytium to crawl upon in order to reach infected areas of the brain (Tambuyzer et al. give a more detailed insight into the potential of microglia

[Tambuyzer et al., 2008]). This property explains their high grade of motility, which always involves a high grade of activated GTPases [Ridley, 2001]. Thus, for the investigation of RhoA and Rac1 activation in astrocytes, it was necessary to reduce the number of microglia in cell culture by shaking them off the astrocytic layer and removing the microglia-containing medium. Nevertheless, a high number of those cells remained among the astrocytes. This had to be taken into consideration by evaluation and interpretation of the results of pull down assays (Fig. 35 A), as these experiments produced highly variable results. Furthermore, it had to be considered that the activation of Rac and Rho in cells is strictly regulated. Upregulation of Rac leads to downregulation of Rho and *vice versa* (as indicated in [Wong et al., 2000] and [Sander et al., 1999]). This further complicates the measurement of activated Rho family proteins. Finally, the compartmentalization of the GTPase response remains largely impossible to analyze.

To approach this problem, and to get further confirmation of the results of the pull down assays, a specific Rac1 inhibitor and a general Rho inhibitor were applied to the astrocytes and the cells were imaged, whereat the observed effects could be directly assigned to the astrocytes. Both approaches, however, showed, that Rac1 and RhoA were not involved in the primary effects (within 5 minutes) induced by sub-lytic concentrations of PLY. The pull down assays showed Rac and Rho activation at 5 minutes, with a maximal activation after 15 minutes. The cellular changes were already observed after a few seconds upon toxin application, and the inhibitors had no capacity to reduce the displacement of cells in the early stages. These findings and the very rapid occurrence of changes indicate that both GTPases more likely play a secondary role in the effects induced by PLY. The Rac inhibitor reduced the displacement of cells in the late phase between 5 and 30 minutes, which is in accordance with Rac1 activation after 5 minutes displayed by the activation assays. The relatively late activation of Rac1 and RhoA thus indicates that, if there is no primary involvement, the late activation could display the attempt of the cell to re-establish the proper functioning of the cytoskeleton. If the toxin leads to a certain cytoskeletal disturbance, the cell will use its capacities to restore the original conditions. In cardiomyocytes, mechanical stretching, which also includes an interference with cytoskeletal integrity, induces Rac1 and RhoA activation in an actin-dependent manner, implying a link between primary actin alteration and secondary GTPase activation [Kawamura et al., 2003]. Alternatively, small GTPase activation by PLY could occur independently of actin remodeling, e. g. via lipid raft interference. Lipid rafts seem to be critically important for proper targeting and activation of small GTPases [del Pozo et al., 2004]. PLY interference with membrane cholesterol (e.g. clustering) [Bonev et al., 2000; Iliev et al., 2007] might additionally modulate small GTPases.

The rather late activation of Rac1 and RhoA, following the rapid effects induced by PLY, implies a primary independence on these signaling pathways and thus points to a direct mechanism of action of pneumolysin on the actin cytoskeleton.

In this context, PLY could possibly interact either with actin directly or through the activation or inhibition of (an) actin effector(s), or through interacting with both actin and an effector in an interconnecting manner.

A certain affinity of the toxin to actin became clear in live cell imaging experiments, which displayed Atto488-tagged PLY, positioned on the membranes of cells in intimate proximity to the actin bundles below the PM. This phenomenon of actin bundle decoration was found at the very early time points after PLY application (within the first 30-60 seconds), and could mainly be observed on top of the cells, whereas lateral binding led to rapid cortex collapse induced by biomechanical tensions (as described above).

Biochemically, the interaction between actin and a possible actin-binding protein can be investigated with the help of co-sedimentation assays. PLY could be detected in the pellet fraction together with actin. The presence of the toxin in the pellet, however, indicated a direct binding to the actin filaments. Furthermore, all variants of PLY had the capacity to co-settle with actin, whereat the wild-type toxin showed the strongest co-sedimentation, and revealed a capacity to slightly stabilize or polymerize actin, indicated by a shift of the actin ratio in direction of F-actin (Fig. 38). Slight stabilization/polymerization of actin could also be observed in actin-pyrene assays, when actin was incubated together with pneumolysin in absence of any other actin-polymerizing factors (Fig. 39). These findings clearly pointed to a direct influence of PLY on actin, confirmed by the fact that the actin polymerization was increased by the toxin in a dose-dependent manner (Fig. 39).

Stabilization of actin includes the establishment of an equilibrium, in which actin monomers assemble and disassemble from a certain filament population in a similar ratio [Alberts et al., 2008]. This indicates that PLY has either an influence on the pointed end of actin filaments where the monomers dissociate, or on the barbed end where the monomers are added to the filament. In the pyrene assays, stabilization is indicated by an increase of fluorescence, with a slightly higher maximal fluorescence rate than the fluorescence observed when actin nucleates spontaneously. The presence of ATP and magnesium is prerequisite for spontaneous dimerization of actin monomers (Fig. 7). The probability, however, that these dimers bind a third monomer, which is the critical step leading to polymerization, is lower than the probability that the dimer dissociates again. The binding of a third monomer increases the probability of successful nucleation and polymerization. The stabilization of actin dimers could be a possible point of attack for PLY to influence actin treadmilling [Alberts et al., 2008].

Pneumolysin settled together with the cytoskeletal components of astrocytes in a cytoskeleton isolation assay. Interestingly, when actin or tubulin, or both components were depolymerized by chytochalsin D, or nocodazole, respectively, the amount of PLY in the pellet decreased, additionally pointing to a capacity of pneumolysin to bind actin filaments, and eventually tubulin structures, too.

Actin affinity was also preserved when actin was partly degraded, indicated by the binding of PLY to actin in Far Western Blotting (overlay blot) experiments (Fig. 40) [Alberts et al., 2008; Hall, 2004]. The capacity to bind not only to actin filaments (F-actin), but also to monomeric actin (G-actin) was displayed by enzyme-linked sorbent assays (ELSAs) (Fig. 41).

Even though this compilation of findings clearly demonstrates the capacity of PLY to interact with actin, both in cells, and within a biochemical background, it appears unlikely that a stabilization or slightly polymerization capacity would be sufficient to lead to the strong cellular modifications induced by the toxin.

In vivo and *in vitro,* the VCA-domain of WASP family proteins is a critical activator of the Arp complex, which in turn is a very potent activator of *de novo* actin nucleation. VCA-activated Arp induces a fast and strong actin polymerization [Machesky and Gould, 1999; Rohatgi et al., 1999].

Actin-pyrene polymerization assays revealed a clear potential of PLY to enhance actin polymerization through the activation of Arp, even in the absence of VCA (Fig. 46, Fig. 48). This finding implied that both proteins are able to interact with each other. Similar experimental setups that were used to investigate PLY-actin interaction (ELSA, imaging of fluorescently labeled proteins) showed that pneumolysin was able to potently bind Arp2 of the Arp complex, and that PLY and Arp2 co-localized in primary astrocytes in close proximity, and partly overlapping (Fig. 43, Fig. 44). Arp is a stable heptameric protein complex, thus revealing multiple binding-sites for different interaction partners [Kelleher et al., 1995; Mullins et al., 1997]. While Arp2 [Rouiller et al., 2008] or Arp3 [Boczkowska et al., 2008] bind to VCA and actin, PLY had high affinity to Arp2, which suggests a bridging function of the toxin between Arp2 and actin. This in turn could explain the actin polymerization capacity of PLY in presence of Arp and absence of VCA (Fig. 46).

The imaging of actin-rhodamine filaments revealed a similar actin phenotype, when Arp2/3 was activated with PLY and with VCA. This additionally indicates, that PLY can activate Arp in a VCA-like manner, inducing the typical actin branching, that is also seen in the lamellipodia of moving cells, indicated in Fig. 12 [Heasman and Ridley, 2008].

Additionally, in the presence of VCA, PLY led to an enhanced actin nucleation and polymerization (Fig. 47), which indicated cumulative effects of the two components, and suggests that PLY binds to a different Arp subunit as VCA, instead of competing for the same binding site (indicated in Fig. 55).

In the context of actin remodeling effects and active cell motility, in which the cells must direct actin assembly with a high degree of spatial and temporal resolution in response to extracellular signals, Cdc42 (another member of the Rho protein family) is suggested to be linked to actin polymerization over N-WASP and Arp2/3 [Rohatgi et al., 1999]. However, it has been shown before that pneumolysin does not increase activation levels of Cdc42 in neuroblastoma cells [Iliev et al., 2007], nor could an activation of this Rho protein be observed in primary astrocytes (data not shown). The absence of Cdc42 in other experimental setups (GUV approach) further confirmed the independence of effects from this protein.

Bacterial protein toxins are widely known to target the actin cytoskeleton, either through direct modification of actin (polymerization, stabilization, nucleation, filament bundling, ADP-ribosylation of specific actin residues) or through interaction with actin effectors (Rho proteins, Arp2/3) (reviewed in [Barbieri et al., 2002] and [Aktories et al., 2011]). Several bacterial effectors were found to activate the Arp complex und thus actively enhance actin polymerization, both *in vitro* and *in vivo*. ActA, a surface protein of *Listeria monocytogenes* is a very potent activator of Arp and leads to the formation of so called comet tails or actin clouds in the cytosol of the host cells. The bacteria keep associated to the built-up actin structures [Welch et al., 1997]. The RickA surface-protein of *Rickettsia conorii* leads to the formation of filopodia-like actin aggregations, similar to ActA via activation of Arp2/3 [Gouin et al., 2004]. Both effectors are surface proteins of obligatory intracellular bacteria. In this respect, they differ from pneumolysin, which represents a virulence factor of an extracellular bacterium and mainly known for its pore-forming properties.

Pneumolysin is not known to be secreted into the host cell. Furthermore it is not known that the actin changes by PLY contribute to an intra-/transcellular movement.

Some evidence exists, however, that pneumococci can be transported transcellularly [Mook-Kanamori et al., 2011]. This study, however, clearly shows that the toxin has the capacity to interact both with actin and Arp2, which are cytosolic proteins. Thus, the question arose if a pore-forming toxin could overcome the membrane barrier in order to exert its functions on the cytosolic side. By using a biomimetic system (GUV approach), composed of a cholesterol-containing lipid-bilayer, actin, Arp2/3, magnesium, ATP and PLY, the question could easily be approached. The setup did not contain any "interfering" components, such as GTPases or VCA. Therefore, direct effects of the toxin could be uncovered. PLY-GFP was delivered to the GUVs with the help of a lipid carrier. The complete free extracellular toxin was inactivated with the help of a cholesterol trap, and the proper orientation of PLY-GFP on the membrane was additionally confirmed with the help of the fluorescent quenching properties of trypan blue. The extra-vesicular suspension contained only the fluorescently labeled actin, the Arp proteins, magnesium and ATP. Actin has the capacity to aggregate spontaneously in the presence of the two latter factors, which could be observed in the suspension containing "blank" GUVs. The actin aggregations around the PLY-GFP loaded GUVs, in contrast, clearly derived from PLY impact on actin, as highest aggregations of actin-rhodamine co-localized with the highest aggregations of PLY-GFP (Fig. 36).

During the pore-formation process, every monomer of PLY unfolds its domain 3, resulting in penetration of the bilayer by four β-sheets of each monomer. In the unfolded conformation, the length of one PLY molecule is more than 12 nm [Tilley et al., 2005]. The thickness of the membrane is about 2.5-7.5 nm. Thus, the unfolded domain 3 could easily interact with the actin cortex below the plasma membrane, or with the Arp complex, which is sequestered and accumulated at the place of actin assembly. This hypothesis was consistent with the findings, that the Δ6 mutant of PLY, which is unable to unfold its domain 3 properly and thus to penetrate the membrane, could not aggregate actin to the same extent as the wild-type toxin. The usage of proper controls (mock, Δ6 mutant) also diminished the possibility that membrane curvature could induce actin aggregation. This had to be ruled out as possible factor, as in the GUVs actin and PLY take opposite positions as in living cells.

Cholesterol clustering by PLY [Bonev et al., 2000] should also be excluded as actin-sequestering factor, because there is no evidence for a direct interaction between actin and cholesterol so far.

As pneumolysin binds to the membrane and is not delivered into the host cell, only the unfolding-domain or parts of it would have the capacity to interact with cytosolic components. Following the finding that pneumolysin has the capacity to overcome the PM barrier by unfolding of domain 3, the dependence on this structural feature needed further investigation. The availability of variants/non-pore forming mutants (W433F [Korchev et al., 1998], Δ6 [Kirkham et al., 2006]) and fragments (D123, D4) of the toxin allows further research on the dependence on pore-formation [Förtsch et al., 2011], and possible binding sites for interaction with other proteins.

Enzyme-linked sorbent assays (ELSAs) revealed decreased binding affinity of the membrane-binding domain 4 and the pore-formation deficient mutant Δ6 both to actin and to Arp2 (Fig. 49, Fig. 50), consistent with the estimation that proper unfolding would be essential for interaction, and that the binding domain would have no binding capacity as its affinity is destined for cholesterol in the membrane [Baba et al., 2001]. The unfolding domain (D123) fragment alone and the WT-toxin, in contrast, biochemically showed high affinity to actin and Arp. These findings strongly point to domain 3 as interaction partner with the intracellular proteins.

The unfolding domain itself consists of six polypeptides, incorporated in the 4 β-sheets that penetrate through the bilayer (Fig. 51). In this work, single peptides were used to reveal possible interaction sites of the domain and actin/Arp. Therefore, the fragments of domain 3 were localized and commercially synthesized (Fig. 51). In actin-pyrene polymerization assays, the application of the single peptides, in addition to PLY, would either diminish or increase actin polymerization effects, either through competition of the peptide with the complete toxin molecule for the binding site within the actin or Arp molecule, or through additive effects, respectively. As none of the peptides led to significant changes of the normal activation grade induced by PLY (Fig. 52), one can speculate that none of the single peptides had the capacity to induce the effects alone. So either multiple segments are needed, or only the complete, properly functioning molecule can induce the effects.

Concluding, these findings clearly reveal domain 3 as the actin/Arp-binding structure, whereat it is not yet clear, if single binding sites in the protein domain exist, or if the domain as a whole is essential for the interaction and the resulting effects.

FRET (fluorescence resonance energy transfer) experiments with PLY-Atto488 and phoalloidin-Alexa555 as FRET pair provided further insight into the structural characteristics of the PLY molecule. Phalloidin is one of the most specific actin binding chemical toxins [Wulf et al., 1979], and has been found to also interact with Arp2 and weaker with Arp3 [Mahaffy and Pollard, 2008]. Considering the simultaneous affinity of PLY towards actin and Arp2, it was interesting to see if there would be phalloidin-specific binding sites on PLY. Indeed, the increase of phalloidin concentration led to quenching of PLY-Atto488 fluorescence, indicative of the presence of FRET. FRET (Fluorescence resonance energy transfer) occurs within a molecular distance of 5 nm, indicative of proximity, and thus coherent with direct interaction [Kenworthy, 2001].

These findings propose that PLY, together with actin and Arp2/3 seem to represent a group of molecules, processing similar and complementary binding surface domains.

In a final experiment, the role of the PLY/Arp interaction on the capability of PLY to form lytic pores was examined. To avoid the interference of cell trafficking phenomena on the toxin's lytic capacity [Idone et al., 2008], we utilized a system that is devoid of endo-/exocytotic events, namely - erythrocyte ghosts (depleted of ATP, and all other cytosolic components, too) and resealed after lytic emptying. The emptying and resealing of the ghosts made cytoskeleton remodeling (energy-dependent phenomenon) impossible, too [Schriei et al., 1975], thus practically limiting completely the cytoskeleton remodeling effects of Arp. The lytic capacity of

PLY in calcein-loaded Arp2/3-enriched ghosts [Prausnitz et al., 1993] was higher than in IgG-enriched or mock ghosts (Fig. 54).

Although the role of other molecules, associated with the membrane as intermediates between PLY and Arp2/3, cannot be excluded completely, the high affinity of the pore-forming domains to Arp2 and the GUV experiments suggest a direct interaction. Earlier works confirm that only a portion of the whole CDC load on the membrane forms lytic pores [Morgan et al., 1994; Morgan et al., 1995], as the exact factors, determining the transition from prepore to pore still remain elusive. The findings point to a possible capacity of PLY to "hijack" the Arp2/3 complex to function as a scaffold, allowing better molecule alignment in pore configuration and thus enhancing the toxin's lytic capacity. The early co-localization between PLY-Atto488 and Arp2 (Fig. 44) supports this concept.

The changes in astrocytic shape and behavior, induced by pneumolysin provide the bacterium certain advantages while gaining access to the brain tissue. The effects of PLY, such as the induction of brain edema, alter the properties of brain tissue and open up spaces that allow deeper penetration into the brain of the host, and also facilitate the entrance of toxic macromolecules. Thus, the investigation of molecular mechanisms of the toxin effects are of major importance [Hupp et al., 2012].

4 CONCLUDING REMARKS

This work had the aim to investigate the molecular initiation mechanisms of cytoskeletal effects induced by disease-relevant concentrations of pneumolysin, based on the findings in neuroblastoma cells, that sub-lytic toxin concentrations led to cell modifications that were based on cytoskeleton remodeling [Iliev et al., 2007; Iliev et al., 2009]. The concentrations used in structural biology studies of PLY, describing the formation of macro pores and lysis, were about 100-fold higher (3-10 mg/ml) and not physiologically relevant [Gilbert et al., 1999; Morgan et al., 1995]. A PLY concentration of 0.2 µg/ml (comprising an equivalent amount of active toxin molecules in comparison to the more cytotoxic first charge of PLY) was determined to be sub-lytic on primary astrocytes in culture.

The application of such amounts of toxin led to rapid cell shape remodeling, associated with shrinking, non-apoptotic membrane blebbing, and development of focal adhesions (FAs) from distinct focal points. The cellular changes resulted in monolayer disruption and cell displacement. All effects could be considered non-lytic, since the observed cells retained intact membranes (as judged by propidium iodide staining). The course of displacement was divided into an early phase with strong displacement, followed by continuing but slower changes.

A very suitable method was established in order to use the displacement as indicator for toxin effects. The cell borders were tracked throughout a time stack with the help of the ImageJ "manual tracking" plugin, and the length of displacement was determined.

Cytoskeletal destruction with cytochalasin D did clearly diminish cell movement upon PLY treatment, the lack of extracellular calcium, however, did not influence the effects, suggesting actin as a possible responsible factor. The small GTPases Rac and Rho were found to be not activated during the early time points of cell movement, and inhibition of the GTPases was not capable of extinction of the effects, proposing that they were not primarily involved in the effects induced by PLY. Late upregulation of Rac and Rho (5-15 minutes after toxin challenge) pointed to a repair mechanism attempt of the cell to re-establish proper cellular function.

A close look (high-resolution z-stack imaging) on the actin cytoskeleton with the help of LifeAct-TagRFP actin staining revealed decoration of cortical actin bundles by PLY. The toxin assembled on the plasma membrane in intimate proximity to actin, and very rapidly led to an actin cortex collapse in direction of the cell core, while the toxin aggregates remained attached to the membrane. The actin collapse occurred exclusively in horizontal direction, where biomechanical tensions are highest.

Several biochemical approaches revealed an affinity of PLY to actin, as most *in vitro* experiments pointed to an actin stabilization capacity of PLY, in contrast to the actin severing/degradation events visible in the imaging approaches. Taking in consideration that even cellular actin effectors, such as gelsolin, on the one hand sever, and thus disassemble actin, and on the other hand polymerize actin, by generation of new polymerization-competent ends, these two PLY-induced effects do not exclude each other. The toxin possibly severs actin at the lateral binding sites of cells, and thus provokes actin polymerization at different sites.

The biomimetic GUV (giant unilamellar vesicle) approach used a limited set of factors, namely the toxin, actin, and Arp, separated by a lipid-bilayer to analyze PLY/actin effects. The wild-type toxin had the capacity to aggregate actin on the opposite site of the membrane in presence of the actin nucleation factor Arp. The amount of actin clusters was significantly higher than the actin aggregations that occurred spontaneously around the vesicles loaded with the non pore-forming mutant Δ6 or around unloaded vesicles. This confirms that PLY had the capacity to overcome the barrier comprised by a lipid-bilayer and was able to exert its effects on the other side, from a membrane-bound state.

Similar to actin, PLY showed a strong affinity to Arp2 of the Arp complex, and similar to actin, the Δ6 mutant and the D4 binding domain showed low binding affinity to the Arp complex protein.

PLY performed slight actin polymerization/stabilization, Arp activation, and synergistic Arp activation in presence of VCA in actin-pyrene polymerization assays.

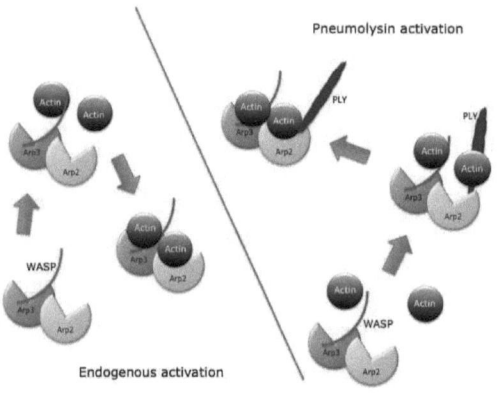

Fig. 55: Schematic of actin nucleation activation.
PLY activates Arp2/3 synergically with the VCA domain of WASP by binding to the Arp2 portion of the complex.

In conclusion, this study revealed novel capacities of pneumolysin that go beyond pore-formation and induction of cell lysis.

The toxin binds to cells, leading to cellular changes that are not associated with cell death. It overcomes the plasma membrane barrier by unfolding its domain 3, and directly interacts with the sub-membranous actin cortex and with Arp, which is accumulated in these actin rich regions.

This study for the first time suggests direct interaction of a toxin of the cholesterol-dependent cytolysin family with submembranous molecular targets.

The effects of sub-lytic amounts of PLY on primary astrocytes, which were uncovered in this work, led to the hypothesis that astrocytic process retraction, cortical astrocytic reorganization, and thus brain tissue remodeling would cause increased tissue permeability to toxic macromolecules and bacteria. This hypothesis was confirmed and additionally revealed an increased interstitial fluid retention in brain tissue, which is manifested as tissue edema [Hupp et al., 2012].

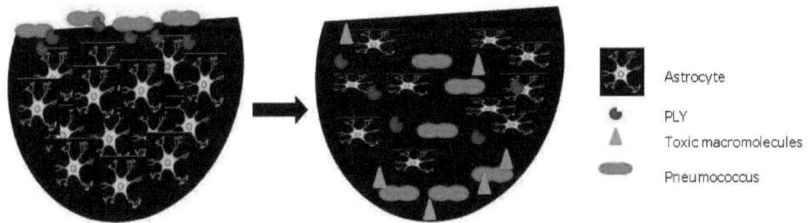

Fig. 56: **Tissue penetration.**
Brain tissue remodeling by astrocytic process retraction and cortical astrocytic reorganization by PLY enables tissue penetration of bacteria and toxic macromolecules.

5 PERSPECTIVES

In general, the investigation of molecular mechanisms of effects induced by bacterial effectors is aimed to get a better understanding of host-pathogen interactions and to possibly discover druggable targets that help to further development of prevention and/or treatment of infections. This work provides a basis for follow-up research in several directions.

Future investigations, based on this work, will have to focus on the mechanistic clarification of PLY-Arp and PLY-actin interaction. As mentioned, this study for the first time proposes that a toxin of the CDC family can act from a membrane-bound state on intracellular actin effector proteins and actin itself. It is indicated here that the unfolding of domain 3 is essential for interaction. However, it has to be further clarified, which peptide structures are responsible for the interaction, and how PLY and the VCA domain collaborate to activate Arp and polymerize actin. Therefore the employment of multiple domain 3 peptides (in different combinations) might give further insight, as possibly not only single peptides, but eventually two or more peptides collaborate to induce actin stabilization or Arp activation, or bridge the interaction of both proteins, respectively.

A closer look has to be taken on the mechanism of actin severing and membrane blebbing, which are possibly linked up. It will be interesting to investigate if the current hypothesis, that cortex severing from the PM and the associated increasing pressure lead to surface distortion, and that the actin stabilization hinders the re-establishment of the cortex, which would induce membrane retraction.

The generation of Arp-knockout astrocytes could further clarify the importance of Arp in the effects induced by the toxin. Literature, however, does not present trials that were successful in knocking out this protein. The cells barely survive this approach. The commercially available Arp2 knock-out cell lines could be applied to gain further insight, although this approach would not sufficiently clarify the effects in context of brain infections.

An animal model will be applied to investigate the potential of pneumolysin to break the blood-brain barrier. Therefore a specific dye will be applied to the tail vein of newborn rats and distribution in the brain after toxin challenge will provide insight into this issue.

Listeriolysin O (LLO) of *Listeria monocytogenes* and perfringolysin O (PFO) of *Clostridium perfringens* belong to the family of CDCs as well. It will be investigated if these toxins, which show high structural homology to PLY, exert the same effects on cells as PLY or if their effects differ from those of PLY. *L. monocytogenes* is not primarily inducing meningitis, but if those bacteria reach the brain, the formation of abscesses can be observed in the sub-meningeal cortex regions. This justifies the application of astrocytes and brain slices for investigation of the effects of LLO. *C. perfringens* is the main cause of food poisoning and gas gangrene. Although not related to meningitis, experimental setups with astrocytes and brain slices will comprise the first approach to gain further insight in the effects of this toxin in cell-based effects.

Abbreviations

µg	Microgram
µl	Microliter
µm	Micrometer
µM	Micromolar
A/Ala	Alanine
APB	Actin Polymerization Buffer
Arp	Actin Related Protein
ADP	Adenosine-5'-diphosphate
ATP	Adenosine-5'-triphosphate
BBB	Blood-brain Barrier
BCA	Bicinchoninic Acid
BL	Black
BSA	Bovine Serum Albumin
C	Centigrade
Ca	Calcium
$CaCl_2$	Calcium chloride
CDC	Cholesterol-dependent Cytolysin

ABBREVIATIONS

Cl	Chloride
CNS	Central Nervous System
CSF	Cerebrospinal Fluid
Cy	Cyanine
CytoD	Cytochalasin D
d, dd	Distilled, double distilled
DABCO	1,4-Diazobicyclo-(2.2.2)-octane
D-MEM	Dulbecco's Modified Eagle Medium
DNA	Deoxyribonucleic Acid
E. coli	*Escherichia coli*
EGTA	Ethylene glycol tetraacetic acid
ELSA	Enzyme-linked Sorbent Assay
EM	Electron Microscopy
F/Phe	Phenylalanine
FCS	Fetal Calf Serum
Fig.	Figure
FITC	Fluorescein isothiocyanate
FRET	Förster/Fluorescence Resonance Energy Transfer

g	Gram
g	Gravitational Force
GAB	General Actin Buffer
GAP	GTPase-activating Protein
GEF	Guanine Nucleotide Exchange Factor
GFP	Green Fluorescent Protein
GDP	Guanosine-5'-diphosphate
GTP	Guanosine-5'-triphosphate
h	Hour
H_2O	Water
HEPES	2-(4-(2-Hydroxyethyl)-1-piperazinyl)-ethanesulfonic acid
HRP	Horse Radish Peroxidase
Ig	Immunoglobulin
K	Potassium
KH_2PO_4	Potassium Dihydrogenphosphate
KCl	Potassium chloride
kDa	Kilodalton
l	Liter

ABBREVIATIONS

LDH	Lactate Dehydrogenase
mA	Milliampere
magn.	Magnification
Mg	Magnesium
mg	Milligram
$MgCl_2$	Magnesium chloride
min	Minute
ml	Milliliter
MOPS	3-(N-morpholino)propanesulfonic acid
mV	Millivolts
n	Sample Size
Na	Sodium
NaCl	Sodium chloride
Na_2HPO_4	Disodium Hydrogenphosphate
ng	Nanogram
NHS	N-hydroxysuccinimid
nm	Nanometer
nM	Nanomolar

ABBREVIATIONS

p	p-Value (probability value)
p (3)	Postnatal day (3)
p(DNA)	Plasmid (DNA)
PAGE	Polyacrylamid Gel Electrophoresis
PBS	Phosphate Buffered Saline
Pen/Strep	Penicillin/Streptomycin
PFA	Paraformaldehyde
PI	Propidium Iodide
PIPES	Piperazine-N,N′-bis(2-ethanesulfonic acid)
PLO	Poly L-ornithine
PLY	Pneumolysin
PM	Plasma Membrane
PVDF	Polyvinylidene fluoride
R/Arg	Arginine
RFP	Red Fluorescent Protein
rpm	Revolutions per Minute
RT	Room Temperature
S.	*Streptococcus*

s/sec.	Second
SD	Sprague Dawley
SDS	Sodium dodecyl sulfate
SEM	Standard Error of the Mean
TMB	3,3',5,5'-Tetramethylbenzidine
TRITC	Tetramethyl Rhodamine Iso-Thiocyanate
V	Volt
v/v	Volume per Volume
VCA	Verprolin-homology-cofilin-homology-acidic (domain)
w/o	Without
W/Trp	Tryptophan
(N)-WASP	(Neuronal) Wiskott-Aldrich-Syndrome Protein
WHO	World Health Organization
WT	Wild-type

References

Abbott, N.J. (2002). Astrocyte-endothelial interactions and blood-brain barrier permeability. *J Anat* 200, 629-638.

Abbott, N.J., Ronnback, L., and Hansson, E. (2006). Astrocyte-endothelial interactions at the blood-brain barrier. *Nat Rev Neurosci* 7, 41-53.

Aktories, K., Ankenbauer, T., Schering, B., and Jakobs, K.H. (1986). ADP-ribosylation of platelet actin by botulinum C2 toxin. *Eur J Biochem* 161, 155-162.

Aktories, K., and Barbieri, J.T. (2005). Bacterial cytotoxins: targeting eukaryotic switches. *Nat Rev Microbiol* 3, 397-410.

Aktories, K., Braun, U., Rosener, S., Just, I., and Hall, A. (1989). The rho gene product expressed in E. coli is a substrate of botulinum ADP-ribosyltransferase C3. *Biochem Biophys Res Commun* 158, 209-213.

Aktories, K., Lang, A.E., Schwan, C., and Mannherz, H.G. (2011). Actin as target for modification by bacterial protein toxins. *FEBS J*.

Alberts, B., Johnson, A., Lewis, J., Raff, M., Roberts, K., and Walter, P. (2008). Molecular Biology of the Cell, 5th Edition edn (Taylor & Francis Group).

Alouf, J.E. (2000). Cholesterol-binding cytolytic protein toxins. *Int J Med Microbiol* 290, 351-356.

Alving, C.R., Habig, W.H., Urban, K.A., and Hardegree, M.C. (1979). Cholesterol-dependent tetanolysin damage to liposomes. *Biochim Biophys Acta* 551, 224-228.

Amdahl, B.M., Rubins, J.B., Daley, C.L., Gilks, C.F., Hopewell, P.C., and Janoff, E.N. (1995). Impaired natural immunity to pneumolysin during human immunodeficiency virus infection in the United States and Africa. *Am J Respir Crit Care Med* 152, 2000-2004.

Araque, A., and Navarrete, M. (2010). Glial cells in neuronal network function. *Philos Trans R Soc Lond B Biol Sci* 365, 2375-2381.

Aspenstrom, P., Fransson, A., and Saras, J. (2004). Rho GTPases have diverse effects on the organization of the actin filament system. *Biochem J* 377, 327-337.

Baba, H., Kawamura, I., Kohda, C., Nomura, T., Ito, Y., Kimoto, T., Watanabe, I., Ichiyama, S., and Mitsuyama, M. (2001). Essential role of domain 4 of pneumolysin from Streptococcus pneumoniae in cytolytic activity as determined by truncated proteins. *Biochem Biophys Res Commun* 281, 37-44.

Barbieri, J.T., Riese, M.J., and Aktories, K. (2002). Bacterial toxins that modify the actin cytoskeleton. *Annu Rev Cell Dev Biol* 18, 315-344.

Barres, B.A. (2008). The mystery and magic of glia: a perspective on their roles in health and disease. *Neuron* 60, 430-440.

Baumann, N., and Pham-Dinh, D. (2001). Biology of oligodendrocyte and myelin in the mammalian central nervous system. *Physiol Rev* 81, 871-927.

Baumgart, T., Hess, S.T., and Webb, W.W. (2003). Imaging coexisting fluid domains in biomembrane models coupling curvature and line tension. *Nature* 425, 821-824.

Berry, A.M., Alexander, J.E., Mitchell, T.J., Andrew, P.W., Hansman, D., and Paton, J.C. (1995). Effect of defined point mutations in the pneumolysin gene on the virulence of Streptococcus pneumoniae. *Infect Immun* 63, 1969-1974.

REFERENCES

Berry, A.M., Lock, R.A., Hansman, D., and Paton, J.C. (1989). Contribution of autolysin to virulence of Streptococcus pneumoniae. *Infect Immun* 57, 2324-2330.

Bingen, E., Levy, C., de la Rocque, F., Boucherat, M., Varon, E., Alonso, J.M., Dabernat, H., Reinert, P., Aujard, Y., and Cohen, R. (2005). Bacterial meningitis in children: a French prospective study. *Clin Infect Dis* 41, 1059-1063.

Bingen, E., Levy, C., Varon, E., de La Rocque, F., Boucherat, M., d'Athis, P., Aujard, Y., and Cohen, R. (2008). Pneumococcal meningitis in the era of pneumococcal conjugate vaccine implementation. *Eur J Clin Microbiol Infect Dis* 27, 191-199.

Boczkowska, M., Rebowski, G., Petoukhov, M.V., Hayes, D.B., Svergun, D.I., and Dominguez, R. (2008). X-ray scattering study of activated Arp2/3 complex with bound actin-WCA. *Structure* 16, 695-704.

Bogaert, D., De Groot, R., and Hermans, P.W. (2004). Streptococcus pneumoniae colonisation: the key to pneumococcal disease. *Lancet Infect Dis* 4, 144-154.

Bohr, V., Paulson, O.B., and Rasmussen, N. (1984). Pneumococcal meningitis. Late neurologic sequelae and features of prognostic impact. *Arch Neurol* 41, 1045-1049.

Bonev, B., Gilbert, R., and Watts, A. (2000). Structural investigations of pneumolysin/lipid complexes. *Mol Membr Biol* 17, 229-235.

Cairns, H., and Russell, D.S. (1946). Cerebral arteritis and phlebitis in pneumococcal meningitis. *J Pathol Bacteriol* 58, 649-665.

Canvin, J.R., Marvin, A.P., Sivakumaran, M., Paton, J.C., Boulnois, G.J., Andrew, P.W., and Mitchell, T.J. (1995). The role of pneumolysin and autolysin in the pathology of pneumonia and septicemia in mice infected with a type 2 pneumococcus. *J Infect Dis* 172, 119-123.

Casella, J.F., Flanagan, M.D., and Lin, S. (1981). Cytochalasin D inhibits actin polymerization and induces depolymerization of actin filaments formed during platelet shape change. *Nature* 293, 302-305.

Charras, G.T. (2008). A short history of blebbing. *J Microsc* 231, 466-478.

Colletier, J.P., Chaize, B., Winterhalter, M., and Fournier, D. (2002). Protein encapsulation in liposomes: efficiency depends on interactions between protein and phospholipid bilayer. *BMC Biotechnol* 2, 9.

Condeelis, J. (2001). How is actin polymerization nucleated in vivo? *Trends Cell Biol* 11, 288-293.

Dagan, R., Givon-Lavi, N., Zamir, O., Sikuler-Cohen, M., Guy, L., Janco, J., Yagupsky, P., and Fraser, D. (2002). Reduction of nasopharyngeal carriage of Streptococcus pneumoniae after administration of a 9-valent pneumococcal conjugate vaccine to toddlers attending day care centers. *J Infect Dis* 185, 927-936.

del Pozo, M.A., Alderson, N.B., Kiosses, W.B., Chiang, H.H., Anderson, R.G., and Schwartz, M.A. (2004). Integrins regulate Rac targeting by internalization of membrane domains. *Science* 303, 839-842.

Eng, L.F., Ghirnikar, R.S., and Lee, Y.L. (2000). Glial fibrillary acidic protein: GFAP-thirty-one years (1969-2000). *Neurochem Res* 25, 1439-1451.

Etienne-Manneville, S., and Hall, A. (2002). Rho GTPases in cell biology. *Nature* 420, 629-635.

Evans, J.H., and Falke, J.J. (2007). Ca2+ influx is an essential component of the positive-feedback loop that maintains leading-edge structure and activity in macrophages. *Proceedings of the National Academy of Sciences of the United States of America* 104, 16176-16181.

REFERENCES

Farrand, A.J., LaChapelle, S., Hotze, E.M., Johnson, A.E., and Tweten, R.K. (2010). Only two amino acids are essential for cytolytic toxin recognition of cholesterol at the membrane surface. *Proc Natl Acad Sci U S A 107*, 4341-4346.

Fiorentini, C., Falzano, L., Travaglione, S., and Fabbri, A. (2003). Hijacking Rho GTPases by protein toxins and apoptosis: molecular strategies of pathogenic bacteria. *Cell Death Differ 10*, 147-152.

Förtsch, C., Hupp, S., Ma, J., Mitchell, T.J., Maier, E., Benz, R., and Iliev, A.I. (2011). Changes in Astrocyte Shape Induced by Sublytic Concentrations of the Cholesterol-Dependent Cytolysin Pneumolysin Still Require Pore-Forming Capacity. *Toxins 3*, 43-62.

Gerber, J., and Nau, R. (2010). Mechanisms of injury in bacterial meningitis. *Current opinion in neurology 23*, 312-318.

Giddings, K.S., Johnson, A.E., and Tweten, R.K. (2003). Redefining cholesterol's role in the mechanism of the cholesterol-dependent cytolysins. *Proc Natl Acad Sci U S A 100*, 11315-11320.

Gilbert, R.J., Jimenez, J.L., Chen, S., Tickle, I.J., Rossjohn, J., Parker, M., Andrew, P.W., and Saibil, H.R. (1999). Two structural transitions in membrane pore formation by pneumolysin, the pore-forming toxin of Streptococcus pneumoniae. *Cell 97*, 647-655.

Gouin, E., Egile, C., Dehoux, P., Villiers, V., Adams, J., Gertler, F., Li, R., and Cossart, P. (2004). The RickA protein of Rickettsia conorii activates the Arp2/3 complex. *Nature 427*, 457-461.

Hall, R.A. (2004). Studying protein-protein interactions via blot overlay or Far Western blot. *Methods Mol Biol 261*, 167-174.

Hartwig, J.H., Bokoch, G.M., Carpenter, C.L., Janmey, P.A., Taylor, L.A., Toker, A., and Stossel, T.P. (1995). Thrombin receptor ligation and activated Rac uncap actin filament barbed ends through phosphoinositide synthesis in permeabilized human platelets. *Cell 82*, 643-653.

Hayward, R.D., and Koronakis, V. (1999). Direct nucleation and bundling of actin by the SipC protein of invasive Salmonella. *EMBO J 18*, 4926-4934.

Heasman, S.J., and Ridley, A.J. (2008). Mammalian Rho GTPases: new insights into their functions from in vivo studies. *Nat Rev Mol Cell Biol 9*, 690-701.

Henrichsen, J. (1995). Six newly recognized types of Streptococcus pneumoniae. *J Clin Microbiol 33*, 2759-2762.

Heuck, A.P., Moe, P.C., and Johnson, B.B. (2010). The cholesterol-dependent cytolysin family of gram-positive bacterial toxins. *Subcell Biochem 51*, 551-577.

Hirst, R.A., Gosai, B., Rutman, A., Guerin, C.J., Nicotera, P., Andrew, P.W., and O'Callaghan, C. (2008). Streptococcus pneumoniae deficient in pneumolysin or autolysin has reduced virulence in meningitis. *J Infect Dis 197*, 744-751.

Hirst, R.A., Kadioglu, A., O'Callaghan, C., and Andrew, P.W. (2004). The role of pneumolysin in pneumococcal pneumonia and meningitis. *Clin Exp Immunol 138*, 195-201.

Hupp, S., Heimeroth, V., Wippel, C., Förtsch, C., Ma, J., Mitchell, T.J., and Iliev, A.I. (2012). Astrocytic tissue remodeling by the meningitis neurotoxin pneumolysin facilitates pathogen tissue penetration and produces interstitial brain edema. *Glia 60*, 137-146.

Iacovache, I., van der Goot, F.G., and Pernot, L. (2008). Pore formation: an ancient yet complex form of attack. *Biochim Biophys Acta 1778*, 1611-1623.

Idone, V., Tam, C., Goss, J.W., Toomre, D., Pypaert, M., and Andrews, N.W. (2008). Repair of injured plasma membrane by rapid Ca^{2+}-dependent endocytosis. *J Cell Biol 180*, 905-914.

REFERENCES

Iliev, A.I., Djannatian, J.R., Nau, R., Mitchell, T.J., and Wouters, F.S. (2007). Cholesterol-dependent actin remodeling via RhoA and Rac1 activation by the Streptococcus pneumoniae toxin pneumolysin. *Proc Natl Acad Sci U S A*.

Iliev, A.I., Djannatian, J.R., Opazo, F., Gerber, J., Nau, R., Mitchell, T.J., and Wouters, F.S. (2009). Rapid microtubule bundling and stabilization by the Streptococcus pneumoniae neurotoxin pneumolysin in a cholesterol-dependent, non-lytic and Src-kinase dependent manner inhibits intracellular trafficking. *Mol Microbiol 71*, 461-477.

Jacobs, T., Cima-Cabal, M.D., Darji, A., Mendez, F.J., Vazquez, F., Jacobs, A.A., Shimada, Y., Ohno-Iwashita, Y., Weiss, S., and de los Toyos, J.R. (1999). The conserved undecapeptide shared by thiol-activated cytolysins is involved in membrane binding. *FEBS Lett 459*, 463-466.

Jaffe, A.B., and Hall, A. (2005). Rho GTPases: biochemistry and biology. *Annu Rev Cell Dev Biol 21*, 247-269.

Jefferies, J., Nieminen, L., Kirkham, L.A., Johnston, C., Smith, A., and Mitchell, T.J. (2007). Identification of a secreted cholesterol-dependent cytolysin (mitilysin) from Streptococcus mitis. *J Bacteriol 189*, 627-632.

Just, I., Selzer, J., Wilm, M., von Eichel-Streiber, C., Mann, M., and Aktories, K. (1995). Glucosylation of Rho proteins by Clostridium difficile toxin B. *Nature 375*, 500-503.

Kadioglu, A., Gingles, N.A., Grattan, K., Kerr, A., Mitchell, T.J., and Andrew, P.W. (2000). Host cellular immune response to pneumococcal lung infection in mice. *Infect Immun 68*, 492-501.

Kalman, D., Weiner, O.D., Goosney, D.L., Sedat, J.W., Finlay, B.B., Abo, A., and Bishop, J.M. (1999). Enteropathogenic E. coli acts through WASP and Arp2/3 complex to form actin pedestals. *Nat Cell Biol 1*, 389-391.

Kandel, E.R., Schwartz, J. H., Jessel, T. M., ed. (1996). Neurowissenschaften - Eine Einführung (Spektrum Akademischer Verlag).

Kastenbauer, S., and Pfister, H.W. (2003). Pneumococcal meningitis in adults: spectrum of complications and prognostic factors in a series of 87 cases. *Brain 126*, 1015-1025.

Kawamura, S., Miyamoto, S., and Brown, J.H. (2003). Initiation and transduction of stretch-induced RhoA and Rac1 activation through caveolae: cytoskeletal regulation of ERK translocation. *J Biol Chem 278*, 31111-31117.

Kayser, F.H., Bienz, K. A., Eckert, J., Zinkernagel R. M. (1998). Medizinische Mikrobiologie, Vol 9 (Thieme).

Kelleher, J.F., Atkinson, S.J., and Pollard, T.D. (1995). Sequences, structural models, and cellular localization of the actin-related proteins Arp2 and Arp3 from Acanthamoeba. *J Cell Biol 131*, 385-397.

Kenworthy, A.K. (2001). Imaging protein-protein interactions using fluorescence resonance energy transfer microscopy. *Methods 24*, 289-296.

Kim, K.S. (2003). Pathogenesis of bacterial meningitis: from bacteraemia to neuronal injury. *Nat Rev Neurosci 4*, 376-385.

Kirkham, L.A., Kerr, A.R., Douce, G.R., Paterson, G.K., Dilts, D.A., Liu, D.F., and Mitchell, T.J. (2006). Construction and immunological characterization of a novel nontoxic protective pneumolysin mutant for use in future pneumococcal vaccines. *Infect Immun 74*, 586-593.

Korchev, Y.E., Bashford, C.L., Pederzolli, C., Pasternak, C.A., Morgan, P.J., Andrew, P.W., and Mitchell, T.J. (1998). A conserved tryptophan in pneumolysin is a determinant of the

REFERENCES

characteristics of channels formed by pneumolysin in cells and planar lipid bilayers. *Biochem J 329 (Pt 3)*, 571-577.

Lilic, M., Galkin, V.E., Orlova, A., VanLoock, M.S., Egelman, E.H., and Stebbins, C.E. (2003). Salmonella SipA polymerizes actin by stapling filaments with nonglobular protein arms. *Science 301*, 1918-1921.

Liu, X., Han, Q., Sun, R., and Li, Z. (2008). Dexamethasone regulation of matrix metalloproteinase expression in experimental pneumococcal meningitis. *Brain Res 1207*, 237-243.

Machesky, L.M., and Gould, K.L. (1999). The Arp2/3 complex: a multifunctional actin organizer. *Curr Opin Cell Biol 11*, 117-121.

Mahaffy, R.E., and Pollard, T.D. (2008). Influence of phalloidin on the formation of actin filament branches by Arp2/3 complex. *Biochemistry 47*, 6460-6467.

Malley, R., Henneke, P., Morse, S.C., Cieslewicz, M.J., Lipsitch, M., Thompson, C.M., Kurt-Jones, E., Paton, J.C., Wessels, M.R., and Golenbock, D.T. (2003). Recognition of pneumolysin by Toll-like receptor 4 confers resistance to pneumococcal infection. *Proc Natl Acad Sci U S A 100*, 1966-1971.

Marriott, H.M., and Dockrell, D.H. (2006). Streptococcus pneumoniae: the role of apoptosis in host defense and pathogenesis. *Int J Biochem Cell Biol 38*, 1848-1854.

Marriott, H.M., Mitchell, T.J., and Dockrell, D.H. (2008). Pneumolysin: a double-edged sword during the host-pathogen interaction. *Curr Mol Med 8*, 497-509.

Masuda, M., Betancourt, L., Matsuzawa, T., Kashimoto, T., Takao, T., Shimonishi, Y., and Horiguchi, Y. (2000). Activation of rho through a cross-link with polyamines catalyzed by Bordetella dermonecrotizing toxin. *Embo J 19*, 521-530.

McGhie, E.J., Hayward, R.D., and Koronakis, V. (2001). Cooperation between actin-binding proteins of invasive Salmonella: SipA potentiates SipC nucleation and bundling of actin. *EMBO J 20*, 2131-2139.

McGough, A., Chiu, W., and Way, M. (1998). Determination of the gelsolin binding site on F-actin: implications for severing and capping. *Biophys J 74*, 764-772.

Mitchell, A.M., and Mitchell, T.J. (2010). Streptococcus pneumoniae: virulence factors and variation. *Clin Microbiol Infect 16*, 411-418.

Mitchell, T.J., Walker, J.A., Saunders, F.K., Andrew, P.W., and Boulnois, G.J. (1989). Expression of the pneumolysin gene in Escherichia coli: rapid purification and biological properties. *Biochim Biophys Acta 1007*, 67-72.

Mook-Kanamori, B.B., Geldhoff, M., van der Poll, T., and van de Beek, D. (2011). Pathogenesis and pathophysiology of pneumococcal meningitis. *Clin Microbiol Rev 24*, 557-591.

Morgan, P.J., Hyman, S.C., Byron, O., Andrew, P.W., Mitchell, T.J., and Rowe, A.J. (1994). Modeling the bacterial protein toxin, pneumolysin, in its monomeric and oligomeric form. *J Biol Chem 269*, 25315-25320.

Morgan, P.J., Hyman, S.C., Rowe, A.J., Mitchell, T.J., Andrew, P.W., and Saibil, H.R. (1995). Subunit organisation and symmetry of pore-forming, oligomeric pneumolysin. *FEBS Lett 371*, 77-80.

Mullins, R.D., Heuser, J.A., and Pollard, T.D. (1998). The interaction of Arp2/3 complex with actin: nucleation, high affinity pointed end capping, and formation of branching networks of filaments. *Proc Natl Acad Sci U S A 95*, 6181-6186.

Mullins, R.D., Stafford, W.F., and Pollard, T.D. (1997). Structure, subunit topology, and actin-binding activity of the Arp2/3 complex from Acanthamoeba. *J Cell Biol 136*, 331-343.

Murphy-Ullrich, J.E. (2001). The de-adhesive activity of matricellular proteins: is intermediate cell adhesion an adaptive state? *The Journal of clinical investigation 107*, 785-790.

Nobes, C.D., and Hall, A. (1995). Rho, rac, and cdc42 GTPases regulate the assembly of multimolecular focal complexes associated with actin stress fibers, lamellipodia, and filopodia. *Cell 81*, 53-62.

Paton, J.C., Rowan-Kelly, B., and Ferrante, A. (1984). Activation of human complement by the pneumococcal toxin pneumolysin. *Infect Immun 43*, 1085-1087.

Pfister, H.W., Feiden, W., and Einhaupl, K.M. (1993). Spectrum of complications during bacterial meningitis in adults. Results of a prospective clinical study. *Arch Neurol 50*, 575-581.

Pourati, J., Maniotis, A., Spiegel, D., Schaffer, J.L., Butler, J.P., Fredberg, J.J., Ingber, D.E., Stamenovic, D., and Wang, N. (1998). Is cytoskeletal tension a major determinant of cell deformability in adherent endothelial cells? *The American journal of physiology 274*, C1283-1289.

Prausnitz, M.R., Lau, B.S., Milano, C.D., Conner, S., Langer, R., and Weaver, J.C. (1993). A quantitative study of electroporation showing a plateau in net molecular transport. *Biophys J 65*, 414-422.

Price, L.S., Langeslag, M., ten Klooster, J.P., Hordijk, P.L., Jalink, K., and Collard, J.G. (2003). Calcium signaling regulates translocation and activation of Rac. *J Biol Chem 278*, 39413-39421.

Ramon y Cajal, S. (1909). Hostologie du systeme nerveux de l'homme et des vertebres. *Malione, Paris*.

Rao, J.N., Li, L., Golovina, V.A., Platoshyn, O., Strauch, E.D., Yuan, J.X., and Wang, J.Y. (2001). Ca2+-RhoA signaling pathway required for polyamine-dependent intestinal epithelial cell migration. *Am J Physiol Cell Physiol 280*, C993-1007.

Ridley, A.J. (2001). Rho family proteins: coordinating cell responses. *Trends Cell Biol 11*, 471-477.

Ridley, A.J. (2001). Rho proteins, PI 3-kinases, and monocyte/macrophage motility. *FEBS Lett 498*, 168-171.

Ridley, A.J. (2011). Life at the leading edge. *Cell 145*, 1012-1022.

Ridley, A.J., and Hall, A. (1992). The small GTP-binding protein rho regulates the assembly of focal adhesions and actin stress fibers in response to growth factors. *Cell 70*, 389-399.

Ridley, A.J., Paterson, H.F., Johnston, C.L., Diekmann, D., and Hall, A. (1992). The small GTP-binding protein rac regulates growth factor-induced membrane ruffling. *Cell 70*, 401-410.

Ridley, A.J., Schwartz, M.A., Burridge, K., Firtel, R.A., Ginsberg, M.H., Borisy, G., Parsons, J.T., and Horwitz, A.R. (2003). Cell migration: integrating signals from front to back. *Science 302*, 1704-1709.

Rohatgi, R., Ma, L., Miki, H., Lopez, M., Kirchhausen, T., Takenawa, T., and Kirschner, M.W. (1999). The interaction between N-WASP and the Arp2/3 complex links Cdc42-dependent signals to actin assembly. *Cell 97*, 221-231.

Rossjohn, J., Polekhina, G., Feil, S.C., Morton, C.J., Tweten, R.K., and Parker, M.W. (2007). Structures of perfringolysin O suggest a pathway for activation of cholesterol-dependent cytolysins. *J Mol Biol 367*, 1227-1236.

REFERENCES

Rouiller, I., Xu, X.P., Amann, K.J., Egile, C., Nickell, S., Nicastro, D., Li, R., Pollard, T.D., Volkmann, N., and Hanein, D. (2008). The structural basis of actin filament branching by the Arp2/3 complex. *J Cell Biol 180*, 887-895.

Rubin, J.L., McGarry, L.J., Strutton, D.R., Klugman, K.P., Pelton, S.I., Gilmore, K.E., and Weinstein, M.C. (2010). Public health and economic impact of the 13-valent pneumococcal conjugate vaccine (PCV13) in the United States. *Vaccine*.

Sander, E.E., ten Klooster, J.P., van Delft, S., van der Kammen, R.A., and Collard, J.G. (1999). Rac downregulates Rho activity: reciprocal balance between both GTPases determines cellular morphology and migratory behavior. *J Cell Biol 147*, 1009-1022.

Schmidt, G., Sehr, P., Wilm, M., Selzer, J., Mann, M., and Aktories, K. (1997). Gln 63 of Rho is deamidated by Escherichia coli cytotoxic necrotizing factor-1. *Nature 387*, 725-729.

Schriei, S.L., Bensch, K.G., Johnson, M., and Junga, I. (1975). Energized endocytosis in human erythrocyte ghosts. *J Clin Invest 56*, 8-22.

Schwan, C., Stecher, B., Tzivelekidis, T., van Ham, M., Rohde, M., Hardt, W.D., Wehland, J., and Aktories, K. (2009). Clostridium difficile toxin CDT induces formation of microtubule-based protrusions and increases adherence of bacteria. *PLoS Pathog 5*, e1000626.

Segal, M.a.Z., B. V. (1990). The Blood-Brain Barrier, Amino Acids and Peptides. (Kluwer Academic, Dordrecht, Boston (USA) and London (UK)).

Shao, F., Merritt, P.M., Bao, Z., Innes, R.W., and Dixon, J.E. (2002). A Yersinia effector and a Pseudomonas avirulence protein define a family of cysteine proteases functioning in bacterial pathogenesis. *Cell 109*, 575-588.

Sofroniew, M.V. (2009). Molecular dissection of reactive astrogliosis and glial scar formation. *Trends Neurosci 32*, 638-647.

Sofroniew, M.V., and Vinters, H.V. (2010). Astrocytes: biology and pathology. *Acta Neuropathol 119*, 7-35.

Spreer, A., Kerstan, H., Bottcher, T., Gerber, J., Siemer, A., Zysk, G., Mitchell, T.J., Eiffert, H., and Nau, R. (2003). Reduced release of pneumolysin by Streptococcus pneumoniae in vitro and in vivo after treatment with nonbacteriolytic antibiotics in comparison to ceftriaxone. *Antimicrob Agents Chemother 47*, 2649-2654.

Squire, L.R., Bloom, F. E., McConnell S. K., Roberts, J. L., Spitzer N. C., Zigmond M. J. (2003). Fundamental Neuroscience (Elsevier Science (USA)).

Stringaris, A.K., Geisenhainer, J., Bergmann, F., Balshusemann, C., Lee, U., Zysk, G., Mitchell, T.J., Keller, B.U., Kuhnt, U., Gerber, J., *et al.* (2002). Neurotoxicity of pneumolysin, a major pneumococcal virulence factor, involves calcium influx and depends on activation of p38 mitogen-activated protein kinase. *Neurobiol Dis 11*, 355-368.

Sun, H.Q., Yamamoto, M., Mejillano, M., and Yin, H.L. (1999). Gelsolin, a multifunctional actin regulatory protein. *J Biol Chem 274*, 33179-33182.

Takai, Y., Sasaki, T., and Matozaki, T. (2001). Small GTP-binding proteins. *Physiol Rev 81*, 153-208.

Tambuyzer, B.R., Ponsaerts, P., and Nouwen, E.J. (2008). Microglia: gatekeepers of central nervous system immunology. *J Leukoc Biol*.

Tilley, S.J., Orlova, E.V., Gilbert, R.J., Andrew, P.W., and Saibil, H.R. (2005). Structural basis of pore formation by the bacterial toxin pneumolysin. *Cell 121*, 247-256.

Tweten, R.K. (2005). Cholesterol-dependent cytolysins, a family of versatile pore-forming toxins. *Infect Immun* 73, 6199-6209.

van Rossum, A.M., Lysenko, E.S., and Weiser, J.N. (2005). Host and bacterial factors contributing to the clearance of colonization by Streptococcus pneumoniae in a murine model. *Infect Immun* 73, 7718-7726.

Varnier, A., Kermarrec, F., Blesneac, I., Moreau, C., Liguori, L., Lenormand, J.L., and Picollet-D'hahan, N. (2010). A simple method for the reconstitution of membrane proteins into giant unilamellar vesicles. *The Journal of membrane biology* 233, 85-92.

Welch, M.D., Iwamatsu, A., and Mitchison, T.J. (1997). Actin polymerization is induced by Arp2/3 protein complex at the surface of Listeria monocytogenes. *Nature* 385, 265-269.

WHO (1999). Report of a meeting on priorities for pneumococcal and Haemophilus infl uenzae type b (Hib) vaccine development and introduction. Geneva, February 9–12, 1999.

WHO (2003). Pneumococcal vaccines. *Wkly Epidemiol Rec* 78, 110-119.

WHO (2007). Pneumococcal conjugate vaccine for childhood immunization—WHO position paper. *Wkly Epidemiol Rec* 82, 93-104.

Wippel, C., Fortsch, C., Hupp, S., Maier, E., Benz, R., Ma, J., Mitchell, T.J., and Iliev, A.I. (2011). Extracellular calcium reduction strongly increases the lytic capacity of pneumolysin from streptococcus pneumoniae in brain tissue. *The Journal of infectious diseases* 204, 930-936.

Wong, W.T., Faulkner-Jones, B.E., Sanes, J.R., and Wong, R.O. (2000). Rapid dendritic remodeling in the developing retina: dependence on neurotransmission and reciprocal regulation by Rac and Rho. *J Neurosci* 20, 5024-5036.

Wulf, E., Deboben, A., Bautz, F.A., Faulstich, H., and Wieland, T. (1979). Fluorescent phallotoxin, a tool for the visualization of cellular actin. *Proc Natl Acad Sci U S A* 76, 4498-4502.

Yarbrough, M.L., Li, Y., Kinch, L.N., Grishin, N.V., Ball, H.L., and Orth, K. (2009). AMPylation of Rho GTPases by Vibrio VopS disrupts effector binding and downstream signaling. *Science* 323, 269-272.

Zysk, G., Schneider-Wald, B.K., Hwang, J.H., Bejo, L., Kim, K.S., Mitchell, T.J., Hakenbeck, R., and Heinz, H.P. (2001). Pneumolysin is the main inducer of cytotoxicity to brain microvascular endothelial cells caused by Streptococcus pneumoniae. *Infect Immun* 69, 845-852.

Acknowledgements

I am deeply grateful to Dr. Asparouh Iliev for giving me the opportunity to work in his group, which was a great and challenging experience all along.

Thanks Aspi, for your constant support and encouragement, for your understanding, communication, your awesome ideas, helping hands, and your faith! I highly appreciate(d) you as my supervisor. You're simply the best!

I am also very grateful to Prof. Dr. Roland Benz for supervising my work, for his support and the fruitful discussions and collaborations. His extensive know-how on bacterial, pore forming toxins was highly beneficial. Appreciation also goes to the rest of the "Benz-group", for the support, and for the experiences and joyful moments we shared.

Special gratitude goes to Prof. Timothy Mitchell, who, despite the great distance between Scotland and Germany, agreed to also supervise my PhD. His invitation to Glasgow was both informative and fruitful, and additionally a great experience. His knowledge about pneumolysin was essential and very helpful for our work.

Most high appreciation goes to Alexandra Bohl for excellent technical assistance. Your support, your helping hands, and not least your friendship made lab life so pleasant and easy. Thank you! I will never forget the time we spent together – at and off work...

BIGUP to the rest of the group! Linus and Förtschi, the last four years were unforgettable. Lab work, seminars, workshops, conferences...we worked as a team (always high throughput), had great fun and success, and despite the general prognoses, we had no cat-fights (haha)... I highly appreciate your friendship.

Special thanks to Vera Heimeroth for the support with the tissue penetration experiments!

Jana, Max, Christine and all the others, thanks for a good time!

Thanks also to Prof. Dr. Martin Lohse and the department of pharmacology for support, fruitful discussions and helpful advices.

Nicht zuletzt möchte ich meinen Eltern danken. Ihr habt mir die Möglichkeit, aber auch die Freiheit gegeben, meinen Weg zu gehen. Ihr habt mich in allen Situationen und Entscheidungen unterstützt. Ganz besonderer Dank gilt meiner Mutter, die sich unermüdlich um Liliane gekümmert hat, was mir mein Leben und meine Arbeit um ein Vielfaches einfacher gemacht hat.

Danke auch an meinen kleinen Professor, Lili, dass du so geduldig mit mir warst, obwohl ich so selten Zeit für dich hatte. Luv up!

i want morebooks!

Buy your books fast and straightforward online - at one of world's fastest growing online book stores! Environmentally sound due to Print-on-Demand technologies.

Buy your books online at
www.get-morebooks.com

Kaufen Sie Ihre Bücher schnell und unkompliziert online – auf einer der am schnellsten wachsenden Buchhandelsplattformen weltweit! Dank Print-On-Demand umwelt- und ressourcenschonend produziert.

Bücher schneller online kaufen
www.morebooks.de

 VDM Verlagsservicegesellschaft mbH
Heinrich-Böcking-Str. 6-8   Telefon: +49 681 3720 174   info@vdm-vsg.de
D - 66121 Saarbrücken   Telefax: +49 681 3720 1749   www.vdm-vsg.de

Printed by Books on Demand GmbH, Norderstedt / Germany